SpringerBriefs in Applied Sciences and Technology

AF787074

SpringerBriefs present concise summaries of cutting-edge research and practical applications across a wide spectrum of fields. Featuring compact volumes of 50 to 125 pages, the series covers a range of content from professional to academic.

Typical publications can be:

- A timely report of state-of-the art methods
- An introduction to or a manual for the application of mathematical or computer techniques
- A bridge between new research results, as published in journal articles
- A snapshot of a hot or emerging topic
- An in-depth case study
- A presentation of core concepts that students must understand in order to make independent contributions

SpringerBriefs are characterized by fast, global electronic dissemination, standard publishing contracts, standardized manuscript preparation and formatting guidelines, and expedited production schedules.

On the one hand, **SpringerBriefs in Applied Sciences and Technology** are devoted to the publication of fundamentals and applications within the different classical engineering disciplines as well as in interdisciplinary fields that recently emerged between these areas. On the other hand, as the boundary separating fundamental research and applied technology is more and more dissolving, this series is particularly open to trans-disciplinary topics between fundamental science and engineering.

Indexed by EI-Compendex, SCOPUS and Springerlink.

Ji Wu • Alvaro Giménez

International Cooperation in Space Science

Springer

Science Press
Beijing

Ji Wu
National Space Science Center
Beijing, China

Alvaro Giménez
Spanish Science Research Council (CSIC)
Madrid, Spain

ISSN 2191-530X ISSN 2191-5318 (electronic)
SpringerBriefs in Applied Sciences and Technology
ISBN 978-981-96-8365-9 ISBN 978-981-96-8366-6 (eBook)
https://doi.org/10.1007/978-981-96-8366-6

© China Science Publishing & Media Ltd. (Science Press) 2025

This work is subject to copyright. All rights are solely and exclusively licensed by the Publisher, whether the whole or part of the material is concerned, specifically the rights of reprinting, reuse of illustrations, recitation, broadcasting, reproduction on microfilms or in any other physical way, and transmission or information storage and retrieval, electronic adaptation, computer software, or by similar or dissimilar methodology now known or hereafter developed.
The use of general descriptive names, registered names, trademarks, service marks, etc. in this publication does not imply, even in the absence of a specific statement, that such names are exempt from the relevant protective laws and regulations and therefore free for general use.
The publisher, the authors and the editors are safe to assume that the advice and information in this book are believed to be true and accurate at the date of publication. Neither the publisher nor the authors or the editors give a warranty, expressed or implied, with respect to the material contained herein or for any errors or omissions that may have been made. The publisher remains neutral with regard to jurisdictional claims in published maps and institutional affiliations.

This Springer imprint is published by the registered company Springer Nature Singapore Pte Ltd.
The registered company address is: 152 Beach Road, #21-01/04 Gateway East, Singapore 189721, Singapore

If disposing of this product, please recycle the paper.

Foreword

Science is discovering the unknowns, and space science is discovering the unknowns of the universe we are living in. The objective of the continuous research is to discover ourselves, "the dust of the stars," who we are, where we are coming from, and where we are going.

Space science has started much before we were able to go to space. Humans have always observed the light coming to them on the Earth, developing more and more sophisticated instruments to understand where they were in the immensity they were traveling in. They have accumulated knowledge over the centuries to the benefit of humanity that they have transmitted from generation to generation. Already at that time, politics was not absent from science within a common understanding between masters and scientists, where masters were supporting and attracting scientists close by and scientists were sharing their progress to the benefit of the masters.

Since 1957, the ability to go to space has created a change of dynamics in three steps, providing a significant advantage to the United States and the Soviet Union and their respective scientists in the first step, which they shared with their respective friends in the second step followed by the extension of access to space to more and more countries and scientists. Access to space has developed in situ research, breaking the long-standing border between the Earth and the universe and leading to a fantastic acceleration of discoveries. The globalization of space science combined with the acceleration of discoveries has multiplied the opportunities for cooperation. Science is a "neutral" territory of cooperation even if politics is still very active.

Among those individuals who have dedicated their entire life to space science, my dear colleagues and friends, Ji Wu and Alvaro Giménez, have joined their respective and joint experience to capture the benefits and the difficulties of international cooperation in space science in this book. Their testimony is not based on theory, but based on hands and practicing, long and wide practicing in concrete cooperative projects, in particular within the European Space Agency (ESA) and between ESA and the Chinese Academy of Sciences. Their objective in writing this book is not so much describing the past, but much more helping the future where international cooperation will be even more important than it could have been so far, since the more we know, the more complex are the unknowns, and the more expertise is required to cross the border of the unknowns.

As they demonstrated in this book, there is not one model of cooperation, but as many models as the number of projects designed and implemented in the frame of international cooperation. As a matter of fact, each cooperation is shaped according to the scientific objectives of the project and the capacities (expertise, funding, etc.) required to implement that project. A wide range of models, including mutual interests among the partners, competition of ideas among the partners, transparency and trust among the partners, and a set of criteria elaborated in the book, among which the role of people is key to the access, who are the owner of a project, are the guarantee of its success, and are the quality of the interaction between the partners.

The people involved in the cooperation would not ignore that the world is full of conflicts and antagonists' interests, but driven by their common objectives; they listen to, understand, and respect the constraints of the other partner, making science the "neutral" territory for progress. Progress is therefore not only a scientific dimension but also a human dimension. Ignorance is the biggest threat to humanity, and knowledge is the biggest opportunity, meaning that science is a laboratory of a better future for the Earth.

You will be more acknowledgeable after having read this book, and thus more confident into the future.

Jean-Jacques Dordain
The institution is European Space Agency (ESA)
Paris, France

Preface

Science is driven by an innate human desire to know about the world around us, how nature works, and our place in it. Space provides an excellent tool for this quest, leading to discoveries and breakthroughs, but also contributing to the advancement of technology and the fascination and inspiration of our society. For these reasons, major spacefaring countries and space agencies consider science as an essential element of their space programs. However, scientific achievements are related to fundamental research for the advancement of knowledge, which may not be the main driver of the space industry or even space agencies. Responsibility for scientific research remains within the science community, supported by universities and research organizations. The result is that the funding of space science missions must come from both sides. Once the mission is approved, its management and funding should be taken by the space agencies, while the science data exploitation must be carried out by the scientists within their research organizations. International cooperation in space science missions shows the same feature. The funding and management of missions are generally the responsibility of space agencies, while the cooperation among the scientists to achieve their goals is supported by the corresponding research entities. These mixed roles may generate problems in space science cooperation activities, which are more difficult to handle than those in other scientific fields that may be less involved in complex and expensive technologies that could also be used for commercial applications and even for national defense.

The twenty-first century is experiencing profound changes on a global scale. Competition, conflicts, sanctions, and even regional wars are constantly endangering cooperation. Since space makes big demands on public funds and advanced technologies, it becomes political. In particular, space is one of the know-hows considered to be strategic in the domain of national security. Thus, countries may not have open access to the essential space technologies of another country, hampering cooperative endeavors. International cooperation in space missions is becoming increasingly difficult compared to the last decades of the twentieth century, when the Americans and the Soviets shook hands in space and the construction of the International Space Station was proposed. Many space science missions were then the result of broad international cooperation, such as the Halley Comet encounter campaign. However, the evolving relations and tensions have deteriorated, making international cooperation increasingly complex, even in science. Our view is that international cooperation in space science should be promoted as an opportunity to face these challenges and a way to build more bridges between different countries.

Nowadays, science and technology are developing very fast, and they are, in fact, deeply intertwined. On the one hand, more and more breakthroughs in science depend on the collective effort of advanced big science instruments/facilities utilizing innovative technologies. Nearly 50% of the Nobel Prizes in Physics since the 1990s have been awarded for advancements based on data from large-scale scientific facilities on the ground and also in space. On the other hand, advances in science inspire the development of new technologies for the benefit of humans on the Earth.[1] Space science remains essential in this context, and international cooperation is certainly an important way to promote it further.

During a review of new space science missions of the Chinese Academy of Sciences in the past years, we stressed the importance of international cooperation despite the existing challenges. However, when we look at the literature, there are actually very few references available in this field. We then arrived at the conclusion that a new book had to be written on this topic on the basis of existing references and our own experience when we held management positions in space science in China and in Europe. In the rapidly changing world of space science, we have to consider new actors, evolving relations, including tensions, and also increasing scientific ambitions despite not increasing budgets. In parallel, we promoted the development of an international discussion forum on the same topic within the activities of the International Space Science Institute (ISSI) in Bern. The conclusions of the forum rather focused on the challenges that cooperation entails and ways for the science community to mitigate them, while this book emphasizes the needs and benefits of cooperation together with the impact of new actors, China in particular.

International cooperation in space science is not just a necessity for the development of space science missions, but also an activity that reflects the vision of science as a global endeavor. Independently of the funding source of space science missions, scientists are committed to the advancement of knowledge for the benefit of humanity. Scientific advance is welcomed, wherever it comes from and whoever brings it. Science is international in essence and as "neutral territory" promotes worldwide dialogue.

For the first draft of this book, Ji Wu was responsible for Chaps. 1, 2, 5, and 7, while Alvaro Giménez took Chaps. 3, 4, and 6. Once the first draft was finished, we cross-read each other's chapters and discussed the points that needed to be clarified. We would like to give particular thanks to the ISSI and the International Space Science Institute in Beijing (ISSI-BJ) for providing workspace and support while we worked concurrently in Bern, Switzerland, and Beijing, China, respectively.

[1] https://www.cas.cn/zjs/202111/t20211110_4813461.shtml [2025-02-06].

Finally, we also thank Ms. Pingping Zhu from Science Press, Ms. Lingyan Zhang from Springer, and Mr. Yongjian Xu from the National Space Science Center (NSSC), Chinese Academy of Sciences, for their careful reading and language suggestions. Without their help, this book would not have been published on time and presented to you.

Beijing, China
Madrid, Spain
June 2024

Ji Wu
Alvaro Giménez

Contents

Chapter 1
Introduction

Scientific research is based on experiments and observations, the competition of ideas, peer review, and the sharing of data and results. That is to say, knowledge must be independent of individuals, non-isolated, unbiased, and accessible to worldwide discussion and verification. Therefore, tools to carry out scientific research and data sharing are fundamental for the advancement of knowledge.

Space science includes a variety of research domains. Nowadays, the term refers in general to all areas of science studying space environment or benefiting from experiments and observations carried out in space. This definition of space science has, of course, evolved over the last decades, from the measurement of the environmental conditions of our planet beyond the upper atmosphere to the test and understanding of the laws of physics that underpin the behavior of the universe, the study of its components, and their behavior. Indeed, space offers access to places for in situ measurements, like the environment of the Sun and the Earth, planets, comets, or asteroids, as well as the possibility of bringing samples back. Moreover, the stable space conditions allow for carrying out long, undisturbed, high-precision, and scientific experiments. However, a key advantage leading to the advancement of space science is given by the possibility to perform unperturbed observations of distant objects, free from the effects of the Earth's atmosphere, across the whole electromagnetic spectrum.

1.1 The Rise of Space Science

In October 1957, humankind entered the so-called Space Age with the launch of Sputnik-1 as shown in Fig. 1.1, the first artificial satellite. Spacecraft in orbit allowed humans to explore new regions beyond the atmosphere and offered new windows to observe the universe without the limitations imposed by it. At that time, little was known about the upper atmosphere, the space environment beyond the ionosphere,

© China Science Publishing & Media Ltd. (Science Press) 2025
J. Wu, A. Giménez, *International Cooperation in Space Science*, SpringerBriefs in Applied Sciences and Technology,
https://doi.org/10.1007/978-981-96-8366-6_1

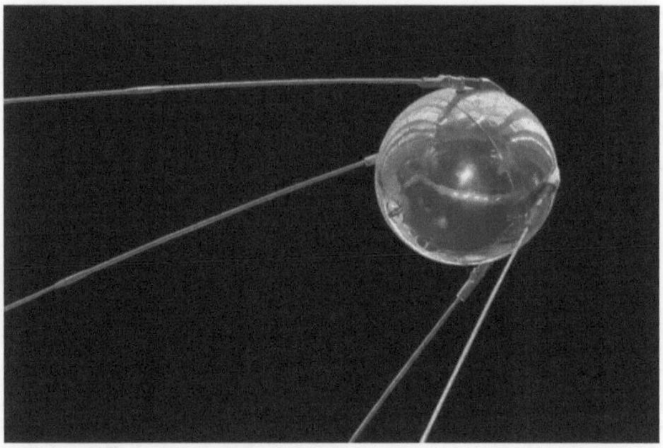

Fig. 1.1 The Soviet Union's first artificial satellite Sputnik 1

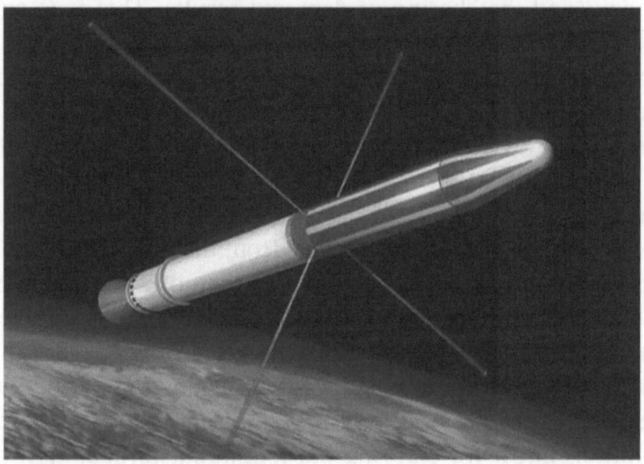

Fig. 1.2 The United States' first satellite Explorer 1

or how cosmic particles penetrate the atmosphere to reach the Earth's surface. In January 1958, Explorer 1, the first satellite of the United States, was launched to explore the radiation environment of geospace as shown in Figs. 1.2 and 1.3. A new research discipline for the study of outer space, known as space research, thus emerged. Moreover, as early as 1958, at the initiative of the United States and the Soviet Union, the International Council for Science Unions (ICSU) formed an ad hoc committee to deal with this emerging field of research, which was not covered by other scientific disciplines at that time. The committee is now the Committee on Space Research (COSPAR), the only international organization in the field of space science and the promoter of exchanges of scientific data among otherwise competing nations.

In the early years of space exploration, a race was unfolding between the United States and the Soviet Union, both interested in leading the advancement in space

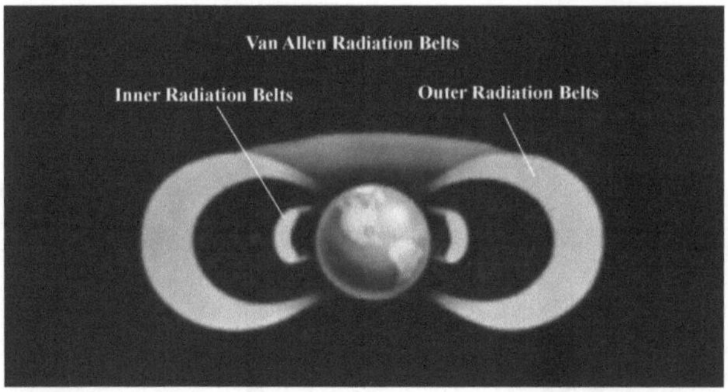

Fig. 1.3 Radiation belts discovered by Explorer 1

and being the first to achieve success in increasingly ambitious challenges. Although most of the efforts were aimed at space technologies, mainly linked to access to space and manned flights in orbit, understanding the space environment was undoubtedly necessary. Consequently, both countries made their best efforts to carry particle and magnetic field detectors, as well as other payloads on board most of their missions for the in situ study of space. As a result, large amounts of measurements of the space environment were obtained, leading to important discoveries, such as the radiation belts. A new discipline emerged with the goal of studying this novel information, which was considered the core of space research and was referred to as space physics or space plasma physics. In the following decades, the study of plasma and its motions in the Earth's environment gradually expanded to encompass the entire interplanetary space. It now covers the study of regions from the upper atmosphere of the Sun to the boundary of the solar system, including the plasma environment around bodies, such as planets and moons.

With a preliminary understanding of the space environment, the opportunity to conduct research from outer space was initiated. In addition to in situ measurements, space provides an excellent platform to observe the stars, and spacecraft were used then to engage in astronomical research. The advantage of space platforms for astronomy is evident. The atmospheric perturbations affecting light from the universe could be easily eliminated from above, providing a new window to observe the universe. The Earth's atmosphere blocks light from reaching the surface in specific wavelength ranges, primarily acting as a barrier to high-energy electromagnetic waves in the ultraviolet, X-ray, and gamma-ray domains. Additionally, there are disruptions in the infrared spectrum. The ionosphere also impacts observations in very-low-radio energy bands. A radio spectrum and its atmosphere windows are shown in Fig. 1.4.

In the 1960s, the United States established three astronomical satellite series, or types of space platforms, for solar, astronomical, and high-energy observations. Thanks to these programs, space astronomy has become an important branch in the field of space science, along with space physics. Nevertheless, unlike space physics, astronomy satellites were often not the only, or even an important, tool to observe the universe. Ground-based telescopes were still the primary source of information

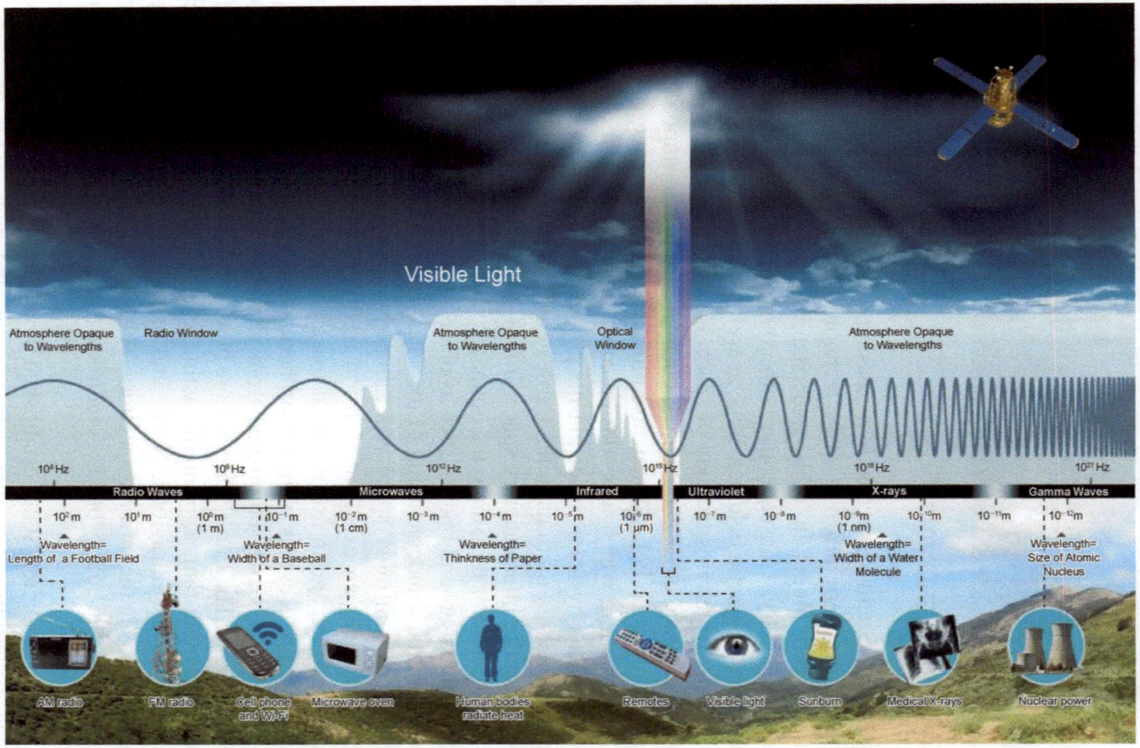

Fig. 1.4 The electromagnetic spectrum and its atmosphere windows

for astronomical research, with space-based observatories serving as a complement, providing an extension of the electromagnetic spectrum beyond the optical domain in the pursuit of understanding the nature and evolution of the universe. Nowadays, even in the optical domain, the highest photometric precision or the homogeneous global scanning of the sky can only be attained from space. In fact, modern astronomy relies heavily on space science.

Almost at the same time, as the early studies of the Earth's space environment, the biggest competition between the United States and the Soviet Union was in deep space exploration, reaching the Moon with manned flights as the priority and the planets with robotic missions. Data were coming in from beyond our planet: the Moon, the Venus, the Mars, and even the planets in the outer solar system. Figure 1.5 shows the first image of the far side of the Moon taken by the Soviet Union's mission Luna 3 on October 1959. Not only space environment data, but more importantly, measurements of the surface, the rocks, and the soil, as well as geological information were obtained through flybys and landings that were very exciting and challenging for researchers. This enabled many Earth atmosphere scientists, geophysicists, and geochemists to join space research, promoting the development of an important area: planetary science.

Unlike the development of other disciplines in space science, the study of the Earth from space began with the advancement of applied remote-sensing satellites. Figure 1.6 shows the first image of the Earth taken from space. The high-altitude location of spacecraft provides a great advantage in observing the Earth's surface. Flying over an orbit hundreds of kilometers high, it is easy to observe and monitor events on the ground within an area spanning thousands of kilometers. Of course, if the weather is such that clouds cover the view, optical sensors cannot penetrate through. However, this issue can be resolved by utilizing radar technologies. Besides, for weather forecasts from space, it is useful to observe the movements and

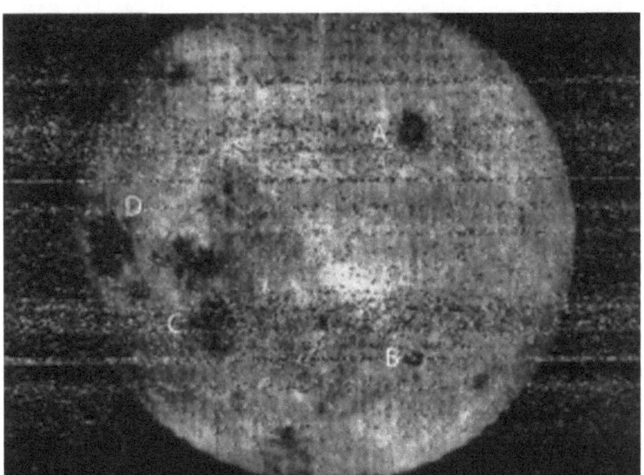

Fig. 1.5 The first image of the far side of the Moon taken by the Soviet Union's mission Luna 3 on October 1959

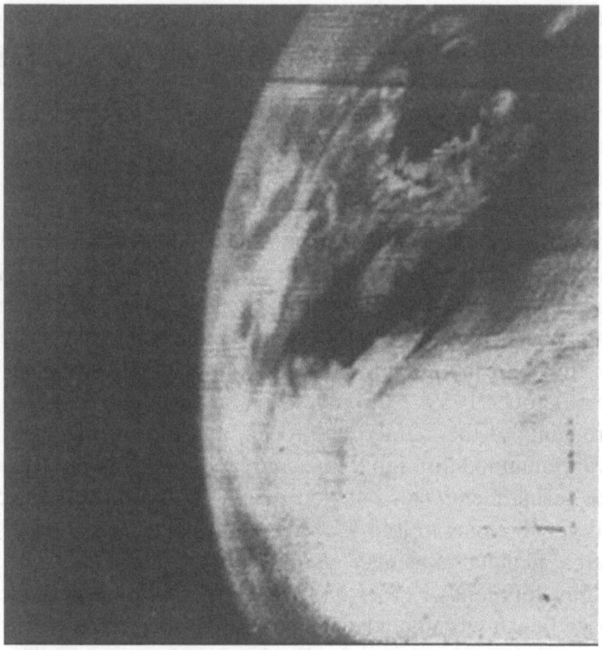

Fig. 1.6 The first image of the Earth taken from space by NASA TIROS-1

Fig. 1.7 Satellite image of a super typhoon in 2014

changes of clouds. Figure 1.7 shows the satellite image of a super typhoon taken from space. Scientists have also utilized ultraviolet light to detect and monitor the ozone hole. Remote-sensing data of the Earth's surface, the oceans, and the atmosphere were gradually accumulated until scientists could study, at the system level, the water cycle of the Earth and that of other chemical components. The study of climate change has facilitated the development of the space Earth science, which has eventually led to the establishment of a new discipline in space science, space

Earth science, based on data provided by satellites orbiting the Earth and observing its behavior from above.

At the beginning of manned spaceflight, new research activities were demanded: microgravity science and space life science. With the challenges of having astronauts in orbit and even flying to the Moon, it was urgent to understand the behavior of materials in microgravity conditions and particularly the human body's exposure to potentially damaging radiation. Scientific research in these two fields had the clear application of enabling manned space missions. However, they are different from other applied space sciences, such as the Earth observation. Remote-sensing images have a very direct mapping function. For example, these images can be used to directly identify targets, such as typhoons, even without a detailed analysis. However, for microgravity condition in space, when the "veil" of the 1 G gravity on the Earth's surface is lifted, what is the physical and biological behavior of matter (especially fluids) and life is a difficult question to answer, requiring important efforts in experiments and research. Figure 1.8 shows the difference between flames on the Earth and in space. The discipline that especially deals with the complex human body system which needs continuous in-depth study is called space medicine. These efforts have promoted the development of microgravity and space life sciences as additional areas of space research. The use of space stations has further allowed the establishment of new experiments in microgravity conditions, including the growth of materials with unique characteristics and the behavior of combustion processes in space. Besides, life science experiments have been expanded to the study of plants and vegetables that could sustain life and the evolution of small animals with short life cycles in conditions that are very different from those on the Earth.

Finally, the special conditions of gravity and vacuum in space allowed for the design of very precise experiments to test essential elements of fundamental

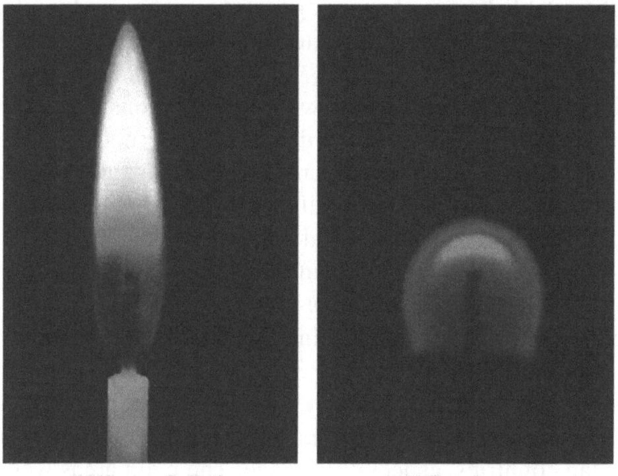

(a) Flame on the Earth (b) Flame in space

Fig. 1.8 Flames on the Earth and in space

physics. The initial studies focused on testing the equivalence principle of general relativity or detecting gravitational waves. More recently, the research has also delved into principles of quantum physics, with potential applications, such as quantum communications. These activities have promoted the field of fundamental physics within space science, and their applications are now being utilized in areas, such as navigation and telecommunications.

In summary, space science encompasses various branches of research, including space physics, space astronomy, planetary science, the Earth science from space, microgravity science, space life sciences, and fundamental physics from space. In other words, space science refers to all possible sciences that can benefit from experiments, measurements, and observations in or from space. Thus, space science can be defined as a cross-disciplinary science using spacecraft, satellites, and probes as tools to study the laws of nature in physics, astronomy, chemistry, and biology across the universe, from the planet Earth to the solar system, the Galaxy, and beyond.

1.2 Actors in Space Science Activities

Tools to advance science have changed from individual curiosity to large-scale projects paid and organized by governments, as is the case of expensive ground and space facilities. International cooperation among scientists is necessary to achieve ambitious goals, but it is not sufficient. It has to be supported by institutional and governmental cooperation among different countries.

Space science is a strategic asset. It ensures technological independence, supports a science-based society and cultural identity, and clearly demonstrates capability and vision. Accordingly, since the very beginning, all activities related to access to space and science were carried out by governments with public funding. This required the establishment of space agencies to define and implement the corresponding programs. The first countries to launch space missions were the Soviet Union and the United States.

In the Soviet Union, space missions were initially considered top-secret military programs, even for those missions that in fact were space exploration missions in planetary science, such as the Lunar, the Venus, and the Mars missions. Changes came with a new institute called the Space Research Institute (IKI) which was established in 1965. The initial purpose of establishing IKI was to have all the research work and the development of science payloads in one single institute and all kinds of tests and facilities under one roof. The new building of this institute was called the "1 km long building" with all kinds of test facilities within it. In the cold winter of Moscow, all instruments could move around from one lab to another without being exposed to open air. IKI also opened flight opportunities to international partners, not only to Eastern European countries but also to Western European countries, especially France. The main areas were high-energy astronomy and the exploration of the Mars with national agencies. However, the European Space Agency (ESA) also benefited from the cooperation experience to launch the International Gamma-Ray Astrophysics Laboratory (INTEGRAL) in 2002 and a mission to the Mars, Mars 96, with European instrumentation, that unfortunately failed but led to the birth of the ESA's Mars Express mission launched in 2003.

In the United States, a new governmental administration was created in 1958 to deal with all space science and exploration missions. This space agency was called the National Aeronautics and Space Administration (NASA). It became the only governmental administration to conduct space science and exploration missions in the United States and, as early as in the first COSPAR meeting in the Hague, NASA offered flight opportunities to foreign payloads and spacecraft. The most famous large program of NASA was the Apollo manned lunar missions, from 1961 to 1972, with six successful landings on the surface of the Moon. Today, NASA is still the world's largest space agency, with the biggest annual budget for space science and exploration.

In Europe, space research missions started at the national level. The UK took the opportunity to cooperate with NASA in a series of small spacecraft called Ariel, mainly providing the science payloads, with Ariel 1 launched in 1962. France established Centre National d'Études Spatiales (CNES) in 1961 and launched its first satellite in 1965. In June 1966, President Charles de Gaulle attended a launch at Baikonur, and shortly after a French–Soviet agreement for cooperation in space science was signed, becoming part of Interkosmos. The gamma-ray mission Signe 3 was the first French-built spacecraft of the cooperation program, launched in 1977. Europe nevertheless decided to advance together with a joint organization called the European Space Research Organisation (ESRO) established in 1964. In 1975, the first successful mission of ESRO was launched on a rocket from the United States, the gamma-ray explorer COS-B. In the same year, ESRO, essentially devoted to space science, was merged with the European Launcher Development Organisation (ELDO), established in the same year as ESRO. The result of the merger was the ESA. The first cooperation of ESA with NASA was the participation of ESA in the originally agreed NASA-UK mission—the International Ultraviolet Explorer (IUE) launched in 1978, providing the solar arrays and a ground station and observing facility, VILSPA, in Spain. ESA started with ten member countries, with the responsibility to conduct most of the European space science and exploration missions, among others. Currently, it has 23 member states, and Canada is an associated member of ESA. The establishment of the ESA nevertheless did not exclude the role of national agencies, which maintained their own programs and bilateral cooperation with non-European partners, as well as their participation in joint activities. This is the case, among others, of CNES in France, Deutsches Zentrum für Luft- und Raumfahrt (DLR) in Germany, and Agenzia Spaziale Italiana (ASI) in Italy. ESA does what individual European nations cannot do on their own.

Japan is another country with an important space science program. Since 1970 when its first satellite was launched, Japan had one science mission almost every year, although the missions were all kept in relatively small size. The institute carrying out these space science missions was the Institute of Space and Astronautical

Science (ISAS), which originated from the Tokyo University. Now, ISAS is part of the Japan Aerospace Exploration Agency (JAXA), but the size and ambitions of the missions have only been continuously increasing.

China and India are two new actors in space science having their first missions only at the beginning of the twenty-first century. China's first space science mission was called the Double Star Program (DSP) which involved strong cooperation with the ESA. India's first space science mission was an astronomical observatory, called AstroSat launched in 2015. In addition, both China and India had ambitious exploration programs. China's Chang'e Program is dedicated to lunar exploration and had already six flights to the Moon. Chang'e-3 and Chang'e-4 achieved soft landings and roving, and Chang'e-4 landed on the far side of the Moon. Chang'e-5 and Chang'e-6 were sample-return missions, and Chang'e-6 had samples being returned from the far side of the Moon for the first time. The first Chinese Mars mission Tianwen-1 is also very relevant. It has the orbiter, lander, and rover all in one mission and landed on the Mars in May 2021. India had three lunar missions. Its first lunar mission, Chandrayaan-1 was launched in 2008. Chandrayaan-2 and Chandrayaan-3 were soft-landing missions and Chandrayaan-3 successfully landed in the South Pole region in 2023. India's Mars Orbiter Mission (MOM) was only an orbiter but arrived at the Mars in 2014, earlier than the Chinese Tianwen-1. Additional new countries are joining the space science club with specific exploration campaigns, such as the Republic of Korea, the United Arab Emirates (UAE), and Israel.

1.3 The Need for International Cooperation

During the first years of space science, competition and national pride allowed fast development and growing budgets. Cooperation in space science was seen as an instrument of geo-politics, which could not be seen as separated from broader aspects of international relations.

Today, cooperation is pursued for a variety of reasons, through a combination of choice and necessity. The leading argument, of course, must be achieving excellent science, but it is obvious that some ambitions cannot be materialized with limited budgets. Moreover, competition may not be affordable at the global level at all times. Compared with other space missions, such as those dealing with commercial applications, science missions provide the best opportunity for international cooperation and allow building bridges for further collaboration. There are many reasons to cooperate in space science among all countries, and these reasons will emerge in the following pages, but we would like to mention the following ones to start with:

1. Science involves a worldwide community.
2. Cooperation maximizes opportunities for excellent science.
3. Science builds cooperation bridges between different countries.

First, because science is a basic aspiration of all countries, new ideas can appear anywhere, and space science cannot afford to miss them due to artificial barriers among countries. Discoveries do not need to be re-discovered, because of mutual denial of competing space missions. Cooperation can avoid overlaps or unnecessary competition in achieving science goals. In this way, the coordinated efforts can complement each other, reaching the "1+1>2" effect, when more or different measurements are available in the space domain. Cooperation accelerates the pace of scientific progress, enables open and shared access to science data, and thus promotes discovery and advancement around the world.

Second, we can mention that the scientific communities behind space science missions are rather mixed. Curiosity has no borders, and one should not do things alone in an isolated manner. Of course, cooperation may also imply sharing the costs of the missions and thus saving part of the adopted budgets for other scientific activities. But it also brings additional costs of coordination and scheduling. A good balance of the cost benefits must be kept in mind, including the science benefits. In this sense, involving more scientists in the data analysis allows for a better science return and increased visibility of the benefits of the investments. Cooperation provides opportunities for shared risks and costs.

Third, the political partnership of countries can be extended through space cooperation. In fact, space science, because of its intrinsic nature of being open worldwide, is able to build bridges and confidence between different countries that would otherwise have difficulties working together. Even for traditional partners in different domains, space science strengthens their cooperation. This is the very essence of ESA with a convention to strengthen space cooperation among its member states, which could be regarded as an international cooperation framework.

The case of ESA is worth some further attention because it comes back to the second half of the last century, when European scientists soon realized that they could not develop and exploit within their own countries the infrastructures needed for world-class science in response to emerging big questions. Only by joining the efforts of several European nations could their advance and leadership be achieved and maintained. In this context, the European Council for Nuclear Research (CERN) was created in 1954 with Edorado Amaldi and Pierre Auger, shown in Fig. 1.9, as key promoting scientists, who were also involved in the new organizations to come. In 1960, the European Preparatory Commission for Space Research (COPERS) started to work, leading to the creation of ESRO in 1964 with the participation of the main European countries. At the same time, in 1962, the European Southern Observatory (ESO) was formed to establish and share new powerful ground-based astronomical facilities in the southern hemisphere (Chile), again with the participation of several European countries. Scientists from European nations can thus function at a world-class level in their specialist fields. The convention of ESRO was ratified in 1964 and it evolved in the following years to finally establish ESA in 1975, including other European space activities but keeping the science program at its core and its only mandatory element. This is why the ESA science program has cooperation in its DNA.

Fig. 1.9 Auger, Amaldi, and Kowarski (from left to right) at CERN in 1952

1.4 Organization of This Book

After a short introduction to the evolution of space science, the actors in space science, and the need for cooperation, mentioned in this chapter, we will discuss in Chap. 2 the fundamental elements in space science programs, such as the funding, where the mission proposals come from and how they are selected, the selection criteria, the scientific instruments' definition, comments about the nature of the space system engineering and in particular its special requirements for science missions, and the key management issues in getting the maximum science output from those missions.

In Chap. 3, we will concentrate on the possible kinds of international cooperation. Different levels of cooperation will involve particular management issues and difficulties. The easiest one is just sharing the data, without any hardware cooperation. On the contrary, the most difficult one is the cooperation with hardware integration into other partners' systems, such as payload on board of partners' platforms or jointly developed satellite platforms. There are also issues of timing to coordinate different development phases. Cooperation may start as early as the planning period, from proposing and selecting a joint mission together and continuing all the way through the mission development.

In Chap. 4, key issues related to international cooperation will be discussed, starting with the science community, a rather special group of people. In most cases, they usually cooperate well with each other, but they do not like top-down directives or work without previous knowledge of the partners, i.e., with the joint group being pushed to work together. Key issues also include the political relationship between collaborating countries and agencies since the funds come from governmental budgets. Trust among the management teams and cultural mutual understanding are very much necessary. If the cooperation is at the system integration level, all issues have to be fully understood and properly taken care of in a satisfactory and agreed manner from the very beginning of the project. Furthermore, we try to display the legal and export restrictions on international cooperation. We start with the most important agreement or memorandum of understanding (MoU) and follow with all other related issues such as intellectual property rights and export restriction regulations, such as the International Traffic in Arms Regulations (ITAR) and the Wolf Amendment passed by the Congress of the United States limiting bilateral space cooperation with China.

In Chap. 5, we will introduce several international organizations that are involved in and may provide support to international cooperation in space science. They are the COSPAR, the International Academy of Astronautics (IAA), the International Astronautical Federation (IAF), and the ISSI. All of them are dedicated to international cooperation by organizing conferences, meetings, bilateral and multilateral summits, joint studies, and publications.

In Chap. 6, we present a few examples of space science missions carried out as international cooperative projects. They represent not only the different types and levels of cooperation but also successful missions and missions that somehow failed, providing lessons to be learned. In these cases, the topics discussed in the previous chapters are further elaborated through the example cases selected.

Finally, Chap. 7 deals with the future of international cooperation, including current science frontiers in which space science communities from all countries are interested. It will also present the fast development of commercial space and its relationship with space science. The fast development of microsatellites and even CubeSats may bring new possibilities for space science using satellite constellations. We will also discuss the advantages and disadvantages of cooperation and competition. At the end, we address the responsibility of the science community in international cooperation endeavors, using their unique position to bring their voice to their governments since science is carried out for the benefit of all human beings.

Chapter 2
Fundamental Elements in Space Science Programs

Spacecraft may serve many purposes, or space applications, such as weather forecast, communications, direct TV broadcasting, land and ocean remote sensing for the survey and management of resources, military intelligence, manned spaceflights, as well as planetary exploration and science. Among all these areas, science research is driving the space science program but also carried out as an important component of manned spaceflight missions and planetary exploration.

Science-related programs have their special features or characteristics. In this chapter, we will discuss the most important ones, which have to be kept in mind by all partners in any international cooperation and by both the science and the management teams. Key elements in science missions are their public funding, the bottom-up character of the selection process, the requirement of excellence of science, the cutting-edge nature of the science instruments, the special requirements for effective system engineering, and the need for an overall maximization of the mission scientific output as an indicator of success.

2.1 Funding

Unlike space programs with commercial goals, space science missions are generally not attractive to private funding though this is very recently changing. The promotion of the understanding of space and the universe, and supporting scientific research, is considered to be a governmental duty in space activities. Therefore, space science is generally supported as part of public spending and managed by governmental space agencies. Some private initiatives have been recently started in the domain of small missions, generally CubeSats, within the framework of the so-called new space approach. They obviously follow different mechanisms in their implementation and exploitation as compared to those discussed in this chapter.

© China Science Publishing & Media Ltd. (Science Press) 2025

J. Wu, A. Giménez, *International Cooperation in Space Science*, SpringerBriefs in Applied Sciences and Technology,
https://doi.org/10.1007/978-981-96-8366-6_2

One important aspect of the public funding of space science programs is that the allocated budgets are based on government decisions about the support provided for the advancement of science and technology. Figure 2.1 shows the limited increase curve of the annual budget of NASA with a great peak only in the Apollo period.[1] The levels of funding are thus limited to an envelope budget within which the space agencies can select and manage specific projects. Unfortunately for the science community, excellent proposals do not print money, thus increasing the available budget above the envelope. They can only be selected if they fit in the available overall budget managed by the agencies carrying out the corresponding programs. It is true that, unfortunately, although scientific excellence is required for each science mission, but not enough to asking more and sometimes even to maintain the neccessary budgets.

Another consequence of this programmatic aspect is that none of the science communities has the privilege to use all of the available budgets. Government agencies need to balance the budgets between different space science sub-disciplines, such as space physics, space astronomy, fundamental physics, planetary science, space Earth science, or microgravity and space life sciences. However, selection criteria should be in favor of the excellence and significance of their science objectives. Therefore, any discipline proposing good science missions, with a big impact in their domain, should have a good chance of being selected.

Besides, the specific science community proposing a project should not consider the mission as their own property. They propose but share ownership with the agencies that ensure the benefits for the overall science community and the science

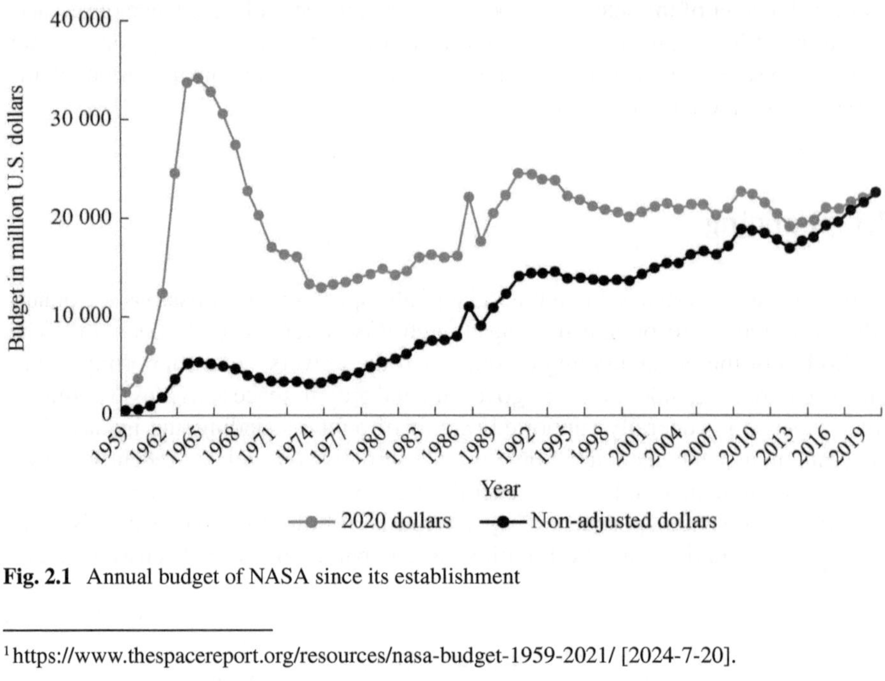

Fig. 2.1 Annual budget of NASA since its establishment

[1] https://www.thespacereport.org/resources/nasa-budget-1959-2021/ [2024-7-20].

program as a whole. Therefore, they must follow the agencies' management and technical constraints, such as proposal selection with well-defined selection criteria, space system engineering development regulations and quality standards, financial and technical feasibility, launcher's capacity, in-orbit operation, and data policies.

2.2 Bottom-Up Selection

Space agencies follow different approaches in the selection of future space science projects. In the case of purely science-driven activities, the bottom-up approach is the strongly preferred process. This allows for a competitive selection of science projects, which are formulated by consortia within the science community in response to a call for the scientific needs of diverse disciplines that can benefit from space measurements and observations. The science community is generally used to the bottom-up and peer-review process and values it as a fair way of making optimum use of the opportunities offered by space programs, as well as in the use of ground-based facilities.

In fact, the proposals for all science programs should come from a bottom-up process. This means that they come from the science communities, the users, not from the governments or agencies' authorities. This is the most fundamental element of space science programs. No matter whether it is about just one science payload on board an application satellite, or a complete mission for a dedicated scientific goal, the initiative of missions in a science program should be coming from the science community. This is not only because most payloads are built by them but also because they are the ultimate users of the data. The principle followed is that agencies provide the tools to do scientific research, but the science community defines the tools they actually need.

There are many positive benefits of the bottom-up approach as follows:

1. Moderated by the peer-review process, the approach ensures that excellent science projects are selected as judged by the best scientists in the community.
2. The projects themselves involve the best scientists, ensuring the scientific relevance and quality of the proposals.
3. The community understands that the procedure is competitive and, once the decisions are made, the community supports the chosen project, resulting in a common sense of purpose.

Nevertheless, there are also some concerns about the bottom-up approach.

1. To reach a consensus, the members of projects have to serve the needs of a broad international science community. Thus, specialized instruments which might be of the highest scientific importance may not be selected, because they serve a too small community. There is a tendency to select multipurpose payloads that will serve broad communities of scientists, making the mission more complex.

Table 2.1 Bottom-up vs. top-down mission selections

Impacts on mission	Bottom-up	Top-down
Initial idea from	Science communities	Agencies
Excellence of science	Guaranteed	Possible
Budget allocation	Competitive	Direct
Science output impact	Larger	Smaller

2. There is a danger that the proposed science objective becomes "conservative" rather than setting ambitious goals which might be riskier.
3. The success in developing new projects strongly depends upon the understanding of the process and the requirements of the funding agencies. Proposers who are newcomers to space science have a very steep learning curve to be successful.

The situation is of course different for application satellites. In this case, there is a much bigger community of institutional users, as well as private developers of commercial services. For example, for weather forecast satellites, the end users are the society at large. The government agency in charge of the weather forecast services takes the role of representing them to propose the program and define the corresponding performance requirements. This is often called a top-down process. On the contrary, the users of scientific research programs are limited to a smaller group of people. This is the research community in a dedicated science discipline, such as the plasma physics community for a particle detection mission or the high-energy astrophysics community for an X-ray observatory mission. Only if the community is interested in the science objectives of the mission, could they have the motivation to analyze and fully exploit the data acquired by the mission. Otherwise, if the program is proposed by the agency in a top-down process, the science community may lose interest or at least will be less motivated to analyze the data, and then the scientific return of the mission will be severely reduced. To maximize the science return is indeed the most important goal for the agency when using public funds for science, and this requires a bottom-up approach. Table 2.1 is a comparison of bottom-up vs. top-down mission selections.

2.3 Scientific Excellence

The purpose of a bottom-up competitive selection process is obviously to achieve scientific excellence with the best possible scientific results. The number of proposals received from the scientific communities is always far larger than the budget available to support them. It is certainly always good to have more ideas than money, rather than the opposite. Selection is, therefore, a necessary process to identify those missions with better potential for a great scientific breakthrough, or larger science impact, including the active involvement of a motivated community. A limited ratio of successful missions, despite leading to a significant number of unhappy scientists, shows a fully competitive selection process, ensures world-class science, and makes the programs sustainable.

The agency in charge should issue a call for proposals with clear selection criteria and a definition of the overall size as the first step of the selection process, in terms of cost to the agency, of the project. After the call is issued, the agency shall carry out a selection of the best candidates out of the submitted proposals. This is a key step in the life cycle of project management, and the agency must ensure the goal of science excellence, while keeping in the selection an open, fair, and unbiased competition.

However, before the scientific aspects are considered, a round of technical and financial feasibility studies should be carried out by the agency's technical offices. This is a sort of screening process to avoid unnecessary efforts in technically non-feasible projects or projects well beyond the budgetary scope of the call. In other words, the science competition should be kept within fair boundaries.

After the screening, a selection focused on the expected science output will then be carried out. The selection criteria of this step shall be released beforehand to let all proposers be aware of them. No matter which agency issues the call, there are always two main aspects in the selection criteria as follows:

1. The impact and ambitions of the scientific objectives, i.e., whether a mission aims at major science challenges and the potential breakthroughs that could change our understanding of the universe and its fundamental laws.
2. The commitment of an excellent science team in achieving the science goals, i.e., whether a mission is supported by a significant number of high-quality scientists involved in analyzing the data acquired using the capabilities of the spacecraft to observe and experiment, producing large amounts of good science as output.

Selected missions should meet in general either one of the two criteria (Wu and Bonnet 2017), and it will increase its success priority if it meets both. In addition to the two above-mentioned selection criteria, the agency also needs to balance the development of different space science disciplines and make sure that the best use of space and ground-based observatories is done in a synergistic way. In the NASA system, priorities and calls are already distributed per discipline, while in ESA all proposals generally compete openly in each opportunity. In this case, when a discipline needs special support, the two criteria could be somehow less strict, reflecting the differences in the corresponding communities.

As discussed above, through the complete selection procedure, the agency shall insist upon two points: first, to ensure that the mission proposals are dealt with a "bottom-up" approach; second, to select the best one according to the above-mentioned criteria. Selection must be done by an independent committee with no conflict of interests; otherwise, a fair and unbiased process would be severely compromised. The outcome of the selection committee at this first stage is generally a limited number of proposals recommended for further study. Since the final decision must be made by the agency issuing the call and responsible for the management and funding of the selected process, the actual technical capabilities available for the science ambitions and the relevant programmatic context have to be further analyzed, in particular "in-kind" contributions and international cooperation. The selected candidates remain in competition during these additional verification studies, and the final selection, or adoption, is eventually done with the independent scientific advice of the agency. In case of large missions requiring the development of technology, long-development periods, or key international partnerships, the selection committee could address a single proposal and the final decision of the agency, or adoption of the mission, is then taken after all the programmatic elements have been successfully addressed.

One may argue that a "top-down" selection could strengthen the leadership of the government agency and thus increase the possibility of having additional funds being allocated. However, it could lead to a situation where the science community may be less motivated to exploit the data and consequently achieve less scientific results. It will also reduce the interest among the scientists to make good new proposals, therefore damaging the competitive process.

From the first round of mission selection, the proposing science teams will have a key role in carrying out the preparatory technical studies, as well as during further phases in the life cycle of the mission. These are the definition and development phases of the mission where the science objectives, and the technical requirements to achieve them, may face all kinds of difficulties, including potential de-scoping exercises. The leader of the proposal is generally called the principal investigator (PI), though his/her role varies in different agencies and types of missions. In all cases, the PI is the key contact point for the agency management structure and the reference in the science community for the full development of the mission and the achievement of its scientific goals.

In different countries and agencies, the management procedure for science projects may change, but the objective is generally the same, to produce excellent science in the end, in terms of breakthroughs, discoveries, and scientific publications. Therefore, for international cooperation, when it involves missions to be defined from the start, selection should be done jointly, and the best way is to issue a single call to the partner countries and set up one selection committee only with the participation of high-level independent scientists from both communities.

2.4 Challenging Scientific Instruments

The scientific method requires multiple experiments and observations with the highest accuracy to be performed. This requirement for the continuous advance of science leads to innovative technology developments representing permanent changes in the embarked instruments on board space science missions. In the case of the Earth, observation programs generally monitor the behavior of the planet, and payloads need stable and repeatable performances to ensure long-period stability of the used data. In contrast, the science mission's objective is discovery, and new measurements and performances are always needed, demanding continuous innovation and change, of course assuming risks. In science, it is generally the case that what has already been done should not be repeated, unless we significantly improve the resolution or the precision of the measurements. The situation is that scientific payloads have to produce new data at each opportunity, and international cooperation is sometimes the way to share cutting-edge technologies for science payloads that could produce new data.

Discovery is to find out events and processes of nature for the first time. The development of a new science mission, aiming at discovery, is thus always linked to or heavily dependent on the development of new science instruments, with unique performances and the best adaptation to specific science goals. They could either have better sensitivity, a larger field of view, a higher space resolution, and a better measurement of rapid time changes or have higher spectral or energy resolution. These improvements are generally the result of an innovative design of the instrument, including optical and thermal elements, and the use of cutting-edge technology, mainly for the detectors.

To develop such challenging instruments, many elements are required, such as materials, detectors, structure designs, mechanisms, optics, computer simulation, and all kinds of environmental tests. In many cases, these elements come from international collaboration teams. However, contrary to commercial procurements, most of the science instruments are developed as in-kind deliveries by the scientific consortia to the funding agency with no money exchange. The party providing the necessary elements to the instrument will benefit by sharing the final science data and the publication of the discoveries and results, not by selling them to other partners. As we will mention in Sect. 4.6, sharing technologies and components may have difficulties beyond the wishes of the scientists involved, and cooperation may be, in the best case, restricted to complete instruments as a contribution to a joint payload. Since the funds for developing the instruments all come in general from public sources, it is necessary to sign an agreement between the participating institutions or universities. This is not only to guarantee the funding needed but also to ensure that all the development phases are fully coordinated. Since there is no money exchange, there is no binding contract, and the only tool between the partners is the signed agreement based on mutual trust and a best-effort basis. The science benefits should also be part of the agreement, in data percentage or visibility in the publications. Of course, in the case of a commercial procurement of part of or even the whole instrument, the science data are only shared with the institution responsible for the purchase, but not with the seller or the provider.

2.5 System Engineering

A space science mission is a large system engineering effort. It consists of several complete systems: the spacecraft, including the service module and the payload, mission ground support, including Tracking, Telemetry, and Command (TT&C)

and data receiving, and science operations, among others. Unfortunately, sometimes scientists are not very familiar with a complete understanding of how system engineering impacts the final scientific output of the mission and how to interact with it, in particular for those who are new in space mission development. The most common mistake they may get into is trying to change fundamental requirements at any time, asking the engineering team for impossible changes that cannot be performed in time or are not even feasible. Another common issue is the delay of the science instrument interface definition, development, and even delivery, leading to cost increases and scheduling problems of the project.

Science operation requirements, including the orbit, the data flow, whether real-time data are needed, the observation window attitude, best and optimum observation opportunities, etc., are essential design requirements. Like in all other space missions, these are the fundamental design input requirements. Once they are implemented into the engineering design, any change may lead to a scenario where the complete system has to be re-designed. It will not only incur cost increases but also cause uncertainties in the feasibility of the project. Therefore, all scientific operational requirements need to be defined as early as possible with an independent validation of the scientific performance requirements.

Once the instruments are designed, all interfaces with the spacecraft platform need to be defined in detail. When they have been defined, all changes need to be reviewed since they may involve changes in other subsystems of the platform, such as mechanical interfaces, power supply, or data handling. The more the project approaches to the end of the development phase, the more difficult it is to make changes and pass the necessary acceptance reviews. In particular, for the flight model development phase, any small change may cause new tests of the whole spacecraft, therefore adding risks and costs to the project. Once the pre-shipment review is over, no change is allowed anymore.

In the case of application satellites, the whole payload design is more or less fixed, and the development is similar from one mission to the next, almost like a series of industrial production. However, for the development of science missions, the users are scientists, and they need to be at the frontier of knowledge, incorporating to the missions the latest advances known at any time. This may lead to a desire to constantly update the science requirements. Therefore, having a common knowledge of the system engineering requirements is necessary for the best science output of the mission with the involvement of both the science and the engineering teams working together. For this purpose, there are two key persons identified in the mission development: PI as the head of the science team and the project manager (PM) as the head of the engineering team. To facilitate the day-to-day work in the different phases of the project, some agencies appoint a project scientist (PS) as a kind of mediator between the two with full knowledge and understanding of the science and engineering worlds. International cooperation partners have to be equally part of the project management structure and the joint science working team and are equally involved in the evolution of the joint project at all levels.

2.6 Maximization of the Science Output

Maximizing the science output of a space science mission is the main goal for any selected project, and there is something to do for this purpose at all phases of its design, development, and operations, Fig. 2.2 shows a space science mission's lifetime flowchart (Wu and Giménez 2020).

As explained previously, during the mission selection phase, there are two major criteria to be considered: the science impact and the community involvement. Therefore, if there is a single international committee to select a joint mission between two agencies, they have to agree first on the selection criteria. Once a cooperation mission is selected, a joint science team should also be established with the participation of scientists from all partners.

During the design phase, the spacecraft and operation requirements have to be fully considered and agreed on. In case that the cooperation involves independent platforms, the role of each in accomplishing the same science objective will be different and so will the requirements. The partners must start an exchange of information at the very early stages of the design phase to have a fully satisfactory common understanding.

During the engineering development phase, i.e., phases C and D, science missions should in general follow the spacecraft development procedures. However, if

Fig. 2.2 Space science mission's lifetime flowchart

there is any specification modification, such as any environmental test failure requiring a significant change of specifications, the joint science team has to be involved in the review and the eventual agreement on a solution.

Compared with application missions, the science missions' most important feature lies in the operation phase. Science missions are operated by the users, i.e., the science team and the community behind them. The involvement of the science team is thus essential from the commissioning phase of both the platform and the science payload. Once the commissioning is over, the right to operate the spacecraft should be formally transferred to the science operation team. If the mission is an international cooperation mission, the leading agency will consider having a mechanism to accommodate the partners' requirements during the operations phase ensuring the successful exploitation of the science data.

2.7 Data Policies

During operations, the key element to maximize the science output is to make the most effective use of the science data, supported by a sound data policy. There are usually two types of science data policies. First, the PI-led team has priority to access the data and exclusive use over a certain period of time. The team has the responsibility to not only analyze the measurements and publish results but also calibrate the data and populate an archive for the rest of the community to eventually access and further exploit the obtained science data. Second, data are shared openly from the very beginning, as soon as verified and calibrated, with the entire science community. Any interested scientist can have access to the data in a completely science-ready and user-friendly format. These policies encourage the science output in different ways.

The first policy focuses on encouraging the project team. The accessibility priority and exclusive rights are rewards for the team's contribution over the years from mission proposal to operation and also reward for their proposed science objectives. In general, researchers proposing science objectives are most eager to use the data and most likely to make major scientific contributions with them. Furthermore, if the quality of the data is not good enough, or not properly calibrated and ready to be used, the interest of scientists outside the proposing team will be affected, and the data utilization and the science output will be seriously compromised. Therefore, exclusive rights for a certain period of time not only help protect the enthusiasm of the mission's science team but also give them time to improve both the data quality and its friendly use for non-team members, thus maximizing both the quantity and quality of the science output.

The second policy aims at enabling more people to access the data. It is most suitable for continuous and stable data produced by survey missions, whose observing targets are constantly changing, with no call for observing time, and can continuously produce science output or even new applications. This is the case of the Earth science space missions and solar monitoring observatories. Because the

Table 2.2 Data policy, PI exclusive, vs. open data

Impacts on mission	PI exclusive	Open data
Science outputs	Less	More
Excellence of science	Yes	Enabled
Responsibility of operations	Strong	None
Reward to the proposal team	Strong	Lower
Adapted to mission type	Exploration	Observatory

measurements of given parameters and targets are constantly changing, science missions produce a large amount of new data every day, which is beyond the processing capability of the mission team. If the data are accessible worldwide, benefits can be maximized by producing a larger amount of science output.

In general, data policies are individually tailored to specific science missions. Usually, the first policy is implemented for a period of time, such as half a year to 1 year. Then, it can be evaluated whether the second policy shall be implemented. How long the data will be kept for exclusive use only is subject to the requirements of output maximization and the nature of the mission and its data. There are special cases of science missions, e.g., almost all the solar physics missions, in which the data will be openly shared from the first day, since the image of the Sun is always changing, and no science team has the capacity to fully analyze all the information. Furthermore, solar images have important relevance to space weather forecast. Therefore, opening them to the public from the first day certainly increases not only the science outputs but also their applications. Table 2.2 compares the differences between PI exclusive and open data policies.

There exists a third kind of data policy for the observatory type of space astronomy missions, where the two cases above are combined. Since the observing time allocation is made from an open call and selected among competing proposals, the selected science team may have the exclusive use of the data about the target they asked for a certain period of time, say 6 months to 1 year. After that period, data should be open to the whole community again.

Concerning international cooperation, data policy is an important element of the mission operations that needs to be agreed upon from the very beginning and fully implemented jointly. Responsibility for the exploitation of the science and the population of a single final data archive has to be agreed upon and recorded in the corresponding MoU.

Chapter 3
Models for International Cooperation in Space Science

Partnerships have a fundamental role in developing innovative, balanced, and affordable programs. They increase the opportunities for scientists across different domains of space science, therefore implying that we get more than the sum of the involved parts. International cooperation in space science can take place in several ways and at different levels, depending on the involvement and responsibilities of the partners, as given by the corresponding task agreements. Cooperation may be established at the level of data sharing, including access to observing time, flight opportunities, the provision of science instruments, the development of joint missions, or programmatic coordination. In this chapter, we will focus on the differences between cooperation models, with particular attention to the management implications.

3.1 Data Sharing

Most space agencies around the world now intend to have an open data policy as long as the experiments on board their science missions allow it. As mentioned in the last chapter, open access to data increases the number and scope of publications, enhancing the global impact of a mission. This applies in general to all science projects, but particularly to astronomy and the Earth science missions. The situation is slightly different for projects designed for a dedicated purpose and by a small group of scientists, for example, in cases of in situ planetary exploration, life sciences, or fundamental physics experiments. On the contrary, for missions observing the Sun or monitoring the Earth science parameters, the information is continuously changing, and data are in general made available with no restriction. In particular, for solar missions, the information provided is also necessary for weather forecasts, and the observational data are accessible immediately after they have been acquired. Due to the large amounts of data obtained every day, the more people analyze them,

© China Science Publishing & Media Ltd. (Science Press) 2025

J. Wu, A. Giménez, *International Cooperation in Space Science*, SpringerBriefs in Applied Sciences and Technology,
https://doi.org/10.1007/978-981-96-8366-6_3

the more benefits we can get for the overall science output. Similarly, for space weather and its understanding, data sharing from missions doing in situ measurements is essential. In all these cases, the only obligation for the data users, when they publish their results, is to acknowledge the space mission providing the data and the used instrumentation.

This is obviously the easiest case of international cooperation in space science. Nevertheless, mission operations must include full pipeline software available together with proper calibration for the data to be accessible in a scientifically useful format. For this purpose, restricted access may be essential to guarantee the quality and homogeneity of the data.

For missions where the data are not open to external users immediately, the PI may have the right to keep them for a certain period, such as, 6 months or 1 year. During this period, the PI and his/her team may have exclusive rights to use the data. However, they could invite other scientists to join the team and use the data during this period. This type of cooperation, or data sharing, generally requires a cooperation agreement between the agencies supporting the corresponding scientific activities, in the form of an MoU between the agencies, which should be signed before the mission is launched. In most cases, the partner scientists are also members of the PI team, as co-PIs, when the mission is proposed. Indeed, co-PIs generally make a significant contribution to the science goals when the mission is proposed but may not be much involved in the instrument hardware development.

Although data sharing itself is a kind of cooperation, it is also an essential element of all other types of cooperation, as discussed in the following sections of this chapter. The final goal of cooperation is to have more science results, and thus, there is no cooperation model that does not consider the science data exploitation.

3.2 Observing Time

In the case of astronomical observatory missions, cooperation is open to international partners in a different way. Space observatories allow the carrying out of scientific projects on specific topics through the measurement of selected targets. Similar to ground-based telescopes, astronomers ask for the exclusive use of the facility for a generally short period of time to point and observe specific targets with a well-defined science objective. The coordination of the Time Allocation Committee (TAC) with the equivalent committees for the use of ground-based facilities for the same science goal is recommended, but terms of such cooperation have to be clearly indicated in the calls.

When the mission is selected, the PIs of the instruments on board the telescope get a part of the initial observing time for the implementation of the proposed science plan or core science program to fulfill the global mission objectives originally proposed. This is part of the so-called guaranteed time in reward for the efforts and time devoted to the development of the mission. It is a percentage of the total time available that decreases with the years in operations while the open time increases.

International cooperation for the guaranteed time is part of the agreements made in the MoU as mentioned in the provision of instruments (see Sect. 3.6). Survey missions do not generally include open-time calls and all the data are made publicly available as soon as they are properly calibrated and are available in a scientifically useful format, ready to be analyzed. However, there may be special types of events that may be reserved for a team selected via an open call.

Nowadays, the situation is even more relevant with the coming into the scene of multi-messenger science, i.e., where scientific research is based on the combination of several sources of information. The combination of measurements in different wavelengths of the electromagnetic spectrum is also connected with ground or space-based measurements of gravitational waves, neutrinos, or energetic particles.

For the open time, the agency managing the science operations of the space facility issues calls to the science community asking for proposals, while forming a TAC to select the best ones that will be included in the mission operation planning. Of course, those enabling the best science return are the ones to be prioritized, but some constraints are generally considered to avoid duplications or affecting the guaranteed time, whose core program should be publicly released in the approved science plan.

It is often the case that those astronomers within the agency participating in the mission development, as well as those in the payload teams, can also make the most competitive proposals and get a big part of the open observing time. Their knowledge of the mission and payload performances is always an advantage. In addition, the proposing teams get their funding for research associated with the data exploitation from their national agency or research institutions. Nevertheless, open time is generally accessible to all nationalities, and time allocation is made on the basis of excellent science as the priority criterion.

3.3 Missions of Opportunity

Another type of cooperation is the provision of relatively small elements to a mission already selected and planned by another agency, the leading partner. This could be specific detectors, parts of the payload instruments, ground station support for the operations, or even a launch opportunity. The agency providing the support has the opportunity to participate in the science of challenging missions that could otherwise not be available. The contribution to the lead agency of the mission is based on a non-exchange of funds and is generally based on the understanding that participation is not a requisite for the selection of the mission itself; otherwise, it would fall into the category of cooperation discussed in later sections of this chapter.

The support in terms of ground stations is quite common, given the need to acquire as much data as possible from the resources in orbit. However, in the case of the provision of launchers, some large missions may have spare launch capacity that could be offered to relatively smaller missions for a ride in exchange for science data. This redundant launch capacity may not be too large, from one hundred to a few hundred kilograms, but for small missions, it could be very useful.

In some cases, the mission of opportunity is discussed between the agencies even during the planning phase. Sharing a dedicated launcher by two partners could be considered, but the cooperation between the involved agencies requires an MoU long before. The timing to develop the two spacecraft is critical to meet the same launch date. Any delay of one of the partners will affect the launch date of the other. However, since the two spacecraft are developed separately, the cooperation is relatively easy, apart from the coordination of the development cycle. Generally, the leading partner offers an opportunity to ride a small spacecraft and will then fix the conditions for acceptance to piggy back on the same launcher. In other words, the country providing the launch opportunity has a stronger voice in making decisions on the launch date, and the user of it has to follow the schedule of the leading agency.

The goal of any mission of opportunity is to have access to more science output for a larger community, and the provision of elements of any kind is made in return for scientific data from the mission. The amount of data shared, or access granted, depends on the actual contribution to the mission. The terms of the science teams' access to data should be clearly detailed in the cooperation agreement between the partner agencies.

3.4 Cooperation Between Programs

Important science frontiers could be of interest to the science community across different countries, and agencies will have to respond to the same requests from their own scientific communities. However, frontier science questions are not only important but also challenging, and the way forward may not be just a single mission implemented by one agency. Here comes an excellent opportunity to establish cooperation at the program level. We have to act together by combining cooperation and competition in which cooperation is for sharing lessons learned, and competition is for defining the best solutions.

The cooperation at the program level means several missions dedicated to the same or similar science-wide objective. Figure 3.1 shows the cooperation between programs. The Mars is a good example. There are more than 40 Mars missions from the United States, Russia, Europe, China, India, and the UAE. Key questions, such as the presence of life, past or extant, still remain unsolved. However, the complementarity of missions could be essential. If different agencies perform in-depth explorations on some of the questions or science objectives, we can advance further in the understanding of the problems together and provide answers accordingly.

Another good example of programs carried out in a cooperative way is the formation of a constellation of satellites to address scientific research with individual elements provided by different agencies, such as a constellation to observe the Sun from different perspective angles and even different distances. With this approach, we can achieve a better understanding of solar activity and predict interplanetary

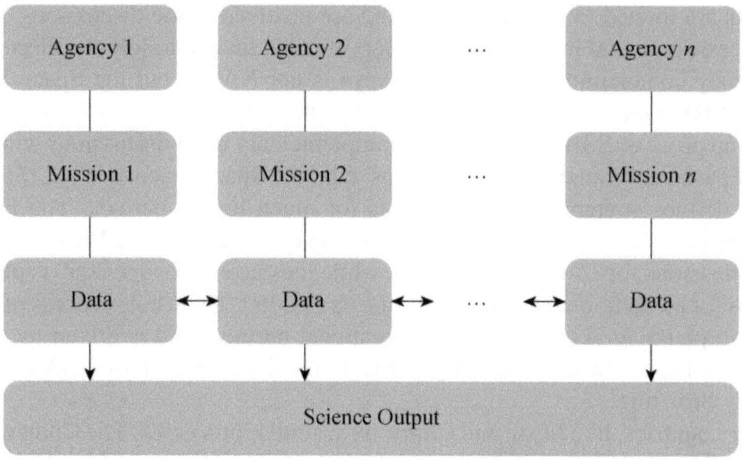

Fig. 3.1 Cooperation between programs

coronal mass ejection (ICME) propagation. This kind of cooperation in the implementation of programs is also a very important type of international cooperation that needs to be planned long before it becomes a reality in space.

Cooperation at the program level starts with the long-term planning activities of different agencies and normally requires that an ad hoc inter-agency group is established. To have program-level cooperation, binding governmental agreements are not needed except for the situation where other actions beyond science are required, as is the case for the International Space Station. Otherwise, agreements at the agency level are needed, particularly to make sure that the timings of the individual efforts, including the launches, are properly synchronized. In many cases, international organizations relevant to space science, like COSPAR, could play an important role in program cooperation plans, helping to set up the above-mentioned ad hoc inter-agency groups and build strategic partnerships.

3.5 Coordination in the Long-Term Strategic Plans

To have long-term cooperation, partners should consider exchanging their views when they are defining their own long-term plans. Large missions defined long in advance provide scaffolding to flexible program elements, develop the necessary technologies, and coalesce partnerships. In addition, sharing long-term plans allows for the development of scientifically complementary missions, rather than unnecessary duplications, increasing the overall science return.

In the United States, plans usually take the form of decadal surveys. They are divided into different science disciplines, such as astrophysics, space weather, planetary science, or earth science. Each of them covers a 10-year vision for the future. The process of carrying out the surveys usually takes 2 years, and hundreds of scientists are invited to send proposals and are involved in the discussions. During this process, potential international partners are also invited to join the discussions. The agency conducting the decadal surveys is not NASA, but the Space Studies Board (SSB) of the National Academies of Sciences. Therefore, it is an independent bottom-up process. SSB also organizes complementary open discussions with international partners through annual meetings, e.g., the Space Science Week (SSW).

ESA defines strategic long-term plans for much longer periods. The Horizon 2000 plan included missions for a period of 20 years. The Cosmic Vision plan considered missions for 15 years up to 2035, while the current Voyage 2050 is planning missions for an additional term of 15 years, up to 2050. The whole process of ESA's long-term planning is also open to international partners and is driven by discussions with the science community through the involvement of ESA's Science Advisory Structure.

Other countries, like Japan and China, have similar processes. The Chinese long-term plan is called Calling Taikong and includes missions up to 2030. Now a new version is being prepared, extending the planning cycle up to 2050.

Cooperation in this long-term planning phase is not about the development of actual missions but about their planning, involving mainly an exchange of information. In the elaboration of these plans, two major issues have to be considered. One is to identify the most important frontiers/challenges of space science. The other is to take care of the sustainability of certain science disciplines. Then the question is how to handle the situation if two agencies are interested in exploring the same frontier or challenge. This is not an academic question. It will always be the case, given the global nature of scientific research. The answer is that international cooperation at the program level will be necessary for the benefit of science in order to improve the use of available resources but mainly to fully exploit the science goals and achievements. We have to look beyond the short term, to make it a preparation for the long term rather than an extension of the past.

3.6 Provision of Instruments

The previously mentioned types of cooperation are relatively easy in the sense that the engineering interfaces can be kept very simple. If we come to even closer cooperation, the first possibility is to provide an instrument to fly on board a partner's mission as part of the scientific payload. In this case, the interfaces between the spacecraft and the instrument are a key element that needs to be carefully considered and agreed upon, since they involve deeper engineering cooperation at the subsystem level. Figure 3.2 shows the cooperation for the provision of instruments.

The motivation for providing instruments as part of cooperation with a partner agency can be as follows:

1. Respond to an open call of the host agency who would like to share their access to space with partners. This is often the case when the host agency is going to fly

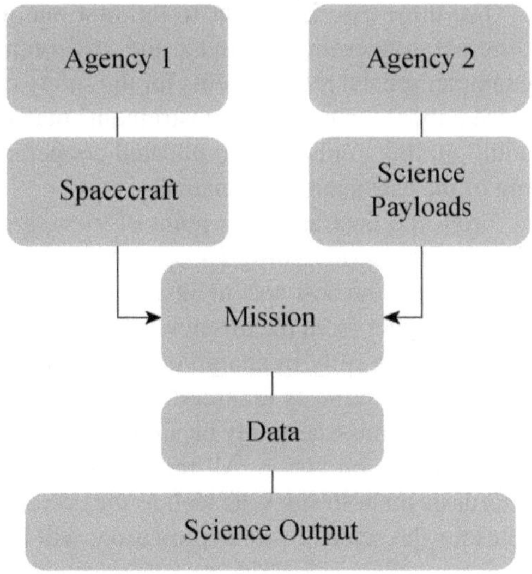

Fig. 3.2 Cooperation for the provision of instruments

to a destination where the other does not have the means or the opportunity to go, like some bodies of the outer solar system.

2. The science team of the host country does not have the capability to build a specific kind of cutting-edge technology instrument, or when the performance of the mission would have much better science output with an additional instrument.
3. The host agency will benefit from fly opportunities on board large platforms, such as space stations.

In the first case, the available resources are already described in the open call, i.e., mass, power, or data download rate. The instrument provider must adapt to the development schedule and all test requirements, having relatively weak control in case of difficulties or changes. In general, they cannot ask for any modification in agreed interfaces or require additional resources. Whether the data will be shared with the host agency depends on specific agreements as part of the mission's science plan. In most cases, the data from the piggy-back instrument are shared with the science teams of the host agency.

In the second case, the instrument provider may have a stronger role due to the key contribution to the overall science objectives. However, all the requirements must be provided at the design phase. Once the design is fixed, there is little chance to make changes. One must adapt to the system engineering and quality control requirements of the host agency. The instrument cannot be treated as a special passenger, but just another element of the whole payload. Data acquired with the provided instrument will certainly be shared with the science teams of the host country and, in most cases, the guest science teams can also access the data from the other instruments onboard. Therefore, this is real science cooperation, as compared with the first case, in which the guest science teams reply to an open call. In other words, this reflects the difference between offering room for an instrument and asking for an instrument that would otherwise not be available.

The third case is similar to the first one, but the requirements are substantially different concerning the mass and environmental and operational conditions. For example, special requirements for the safety of the astronauts/taikonauts are needed, since onboard experiments/instruments may require manned onboard operations. In addition, this implies a complicated cooperation scheme including dedicated training of the astronauts/taikonauts.

From the host agency's point of view, providing an opportunity to the international partner, or issuing an open call, also involves a major responsibility in all three cases. The best way to ensure a smooth development phase is to assign a dedicated engineer from the mission team to work with the instrument science team. He/she should be fully in charge of the accommodation of the guest instrument in the platform and serve as the interface to the cooperation partner agency.

An MoU must certainly be signed between the partner agencies in all three cases of instrument provision. All relevant teams, including the persons in charge of the interfaces on both sides, as well as the corresponding management structure and the roles for the science data exploitation, will be included in the MoU.

3.7 Joint Development of Missions

Probably the most complicated type of international cooperation is to carry out a joint mission, from mission selection to mission development, launch, and science operations. In this case of cooperation, the two agencies involved have to work together very closely from the beginning to the end of the mission. The benefits of the cooperation are also very important because they put valuable resources together, science and engineering, in one mission with joint teams and open exchange of experience and knowledge, thus producing benefits beyond the science goals. Of course, as a normal cooperation endeavor, activities are based on a non-exchange of funds principle, and the stability of the project is based on common goals and benefits with a win–win approach. Figure 3.3 shows the cooperation for the joint development of missions.

The cooperation starts with a joint call for proposals. In order to prepare for the join call, the two agencies could organize workshops together to let the science community present their proposals and discuss their common interests and possibilities for cooperation. Once there are enough ideas for joint mission proposals, the two agencies should issue a joint call to select the best one in terms of first-class science, feasibility, and affordability, through an independent evaluation committee.

As explained in last chapter, the selection process should follow the two key criteria, namely science excellence and community involvement, and it should be conducted by an independent joint committee. This committee should provide a short list of the best proposals with a prioritization for further study. The two agencies will then make their own assessment of the technical feasibility issues of each selected proposal. The results will then be shared and included in the joint assessment of the proposals for their final review. This should be done again by the

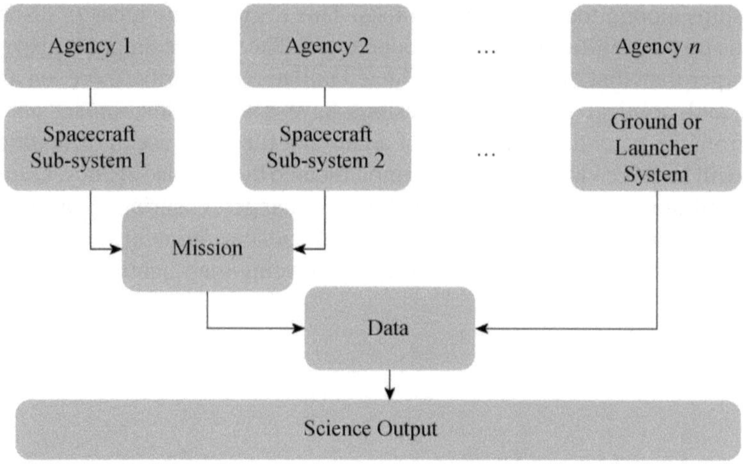

Fig. 3.3 Jointly developed of missions

independent joint committee, and the final decision will then be made by the two agencies together.

In the design phase, the two agencies must agree on the division of labor, i.e., who will take care of what at the system level. For example, if one party takes care of the launcher, the other could take care of the platform development, while the science payloads are normally developed by science teams on both sides. The development phase is probably the most challenging. Since the development schedules could be different for each agency, the coordination and scheduling of joint reviews at each development phase must be agreed upon in detail, both in terms of content and format. The same applies to the different tests and reviews that will be required for the science payloads. Finally, integration and flight acceptance tests will have to be done jointly.

In addition to science, the most important benefit of cooperation in the development of a joint mission is that the experience of the engineering teams could be shared by both parties. Sometimes, one may find the other party's experience inspiring and worth learning from. There is also an opportunity to learn about another way of working during the process of implementing the same procedures, documentation, and reviews throughout the project phases. In the end, these experiences foster long-lasting mutual trust and even friendship among the teams, which is key elements for cooperation.

Finally, science operations will be performed jointly at operation centers on both sides, which also requires working together. The data can be exclusively used by the joint science team first and then open worldwide for further scientific exploitation or fully open from the first day.

This type of cooperation requires a lot of mutual understanding and joint work, but it comes with much larger benefits. Pooling resources not only saves money but also, mainly, fosters connections among engineering teams and strengthens relationships among the involved scientists. This triggers new ideas in discussions and opens up possibilities for further cooperation. The science output is also considerably larger than that of missions conducted unilaterally. Furthermore, on an international and political level, joint missions have a significant impact and could benefit from the potential availability of additional funds in case of difficulties. The mission will be more visible than being implemented by only one agency. Partnership solidifies missions in the face of eventual policy changes, ensuring stability. Finally, when the data are made accessible to the global science community, more scientists will participate in the data analysis, thereby enhancing the scientific output.

Chapter 4
Key Elements for a Successful Cooperation

It is often said that if we want to achieve something fast, we have to do it alone, but if we want to go far, we better do it together. In other words, for space science missions, on the one hand, we may use competition to move faster and cooperation to achieve more. Indeed, cooperation missions involve issues of close coordination that may slow down processes and increase costs, since a significant part of the savings will be used in pooling resources together as well as a more complex development phase. On the other hand, competition enhances the visibility of leadership and the political impact, leading to faster and more efficient processes and even increased budgets. In fact, these general remarks may also be relevant in application programs providing services, but in space science, the benefits of cooperation are much more important than those of competition, of course in the case of successful cooperation, the ultimate goal is to foster cooperation to maximize the science output of a mission that uses public funding in promoting the advance of scientific knowledge.

In this chapter, we discuss possible key elements identified to be necessary to have successful cooperation in a space science program. Success is measured in terms of the final science output of the mission, generally in publications with groundbreaking results. We thus have to consider how to make sure that this is achieved with a good understanding of the role and motivation of each of the actors involved. An essential element is mutual trust, applicable to the science community and the involved space agencies. This is shown by the honest and transparent exchange of all the necessary information for the success of the mission. Very important as well is the commitment and leadership role of the different actors, their responsibility in achieving success, and the understanding of their cultural and national backgrounds. In addition, cooperation is clearly fostered by program stability, shared scientific ambitions, and open data policies.

© China Science Publishing & Media Ltd. (Science Press) 2025

J. Wu, A. Giménez, *International Cooperation in Space Science*, SpringerBriefs in Applied Sciences and Technology,
https://doi.org/10.1007/978-981-96-8366-6_4

4.1 Science Community

A space science mission is a result of the efforts of many actors or stakeholders. They are governmental agencies, science communities, space industries, and the public at large. Figure 4.1 shows the relationship among them. However, the most important in any space science mission is in fact the science community. The relations between the science community and other stakeholders were analyzed by Wu (2022) and briefly shown in Fig. 4.1.

Science must drive any space science project. Whether in cooperation or not, science must go first, and scientists should define the tools they need to achieve and sustain excellence, which must rule choices. The first element for successful cooperation in any space science mission is obviously the capability of a team of scientists to work together in achieving common and ambitious scientific goals. It is also very important that the scientific communities of all partners are fully involved. No partner should be in the situation of enabling the science of the other partner but no science benefit for themselves. Open and efficient cooperation among scientists is a prerequisite for a successful science mission, and the challenge of dealing with what is generally called the "prima donna" syndrome among scientists should not be underestimated. It is important also to consider, in any cooperation scenario, that the science community at large, beyond the proposing team, is the final user, ensuring the full exploitation of its capabilities. The more scientists involve in the data analysis of any science mission, the larger science output will be achieved. In order to obtain the best scientific exploitation of the science mission, an open release of scientific data needs to be a high priority. Of course, this implies that data need to be properly verified and calibrated so that scientists out of the project team can handle the data in the correct scientific format and with user-friendly access tools.

We also have to consider the intrinsic diversity within the science community to achieve the best results. It is important, when selecting new projects, to evaluate not only the size of the involved community, and the number of participating or

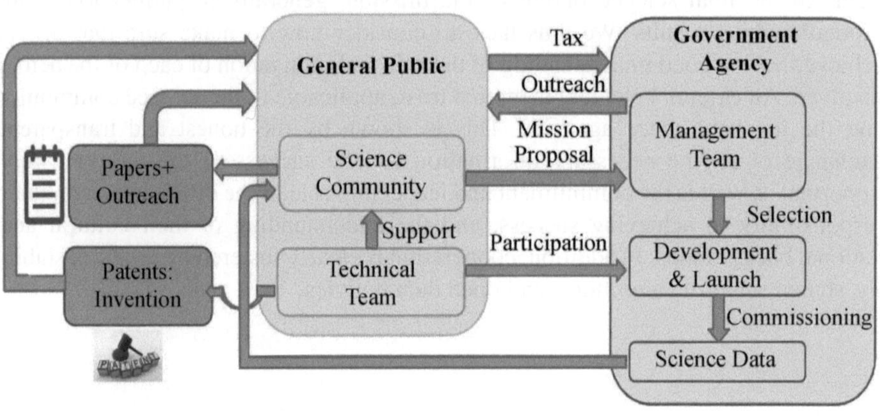

Fig. 4.1 Relationship among the stakeholders of a space science mission

interested scientists, but also how many of them are young, early-career, scientists. This is the only way to ensure that despite the long duration of the project development, the proper exploitation of its scientific outcome is guaranteed. In other words, we have to make sure that new ideas and challenges are envisaged, rather than old routines with no attraction to new young teams, avoiding the sometimes-called welfare science.

Nevertheless, we have to consider new promising and innovative space science disciplines, with initially small communities. This may contradict the selection of projects with a well-established and extensive science community. However, it ensures the progress of space contributions to science, even though it may involve bigger risks. The question to evaluate is the balance between the level of risk and the scientific gain. In contrast, subject matter experts and project leaders may face a "choice overload" with limited capability to fully entertain all partnership opportunities.

Science communities are spread worldwide, across national borders, they are not divided in their quest for scientific answers, and the members often meet in international conferences. However, on the one hand, during the pandemic since 2020, online meetings have also been organized very frequently though they could not replace the benefits of in-person meetings. On the other hand, scientists all over the world are selected to review the publications of other members of the community. It is therefore not difficult in general when a new mission is proposed, to gather a team of top-level scientists to review it. Once a proposal is supported by the community within its science discipline, the scientists who have been involved in the review will make a fundamental contribution to attract a larger group of scientists to join the cooperation in the data analysis. In other words, involving top-level scientists of the relevant discipline community from all over the world at the time of the mission proposal and evaluation phases is essential for the agency management, and it will lay down a solid foundation for possible future international cooperation.

Finally, the importance of outreach involving a given mission should not be underestimated, beyond the science community. Citizens' appreciation may be important for partners in growing space economies to maintain support for public spending. Visibility of the international nature of the collaboration is key to its appreciation in different countries as well as to the long-term stability of the mission development.

4.2 Mutual Trust

Trust among scientists, as mentioned before, is essential and it is expected to be the starting point for any mission project. Moreover, mutual trust is also needed between engineers and agencies for the successful development of a mission and its exploitation. Finally, political and policy support, based on mutual respect, is also necessary for any cooperation to be successful. This support requires the joint project to have scientific and technical merit and demonstrates specific benefits to all partners. Indeed, top-level agency support is necessary for any international cooperation, whether a science mission or not, involving mutual trust which is generally built on the basis of previous experiences, i.e., a good track of successful cases. The reliability of partners is recognized in the fulfillment of earlier commitments.

It has to be clearly understood that good cooperation requires excellent personal relations. Only personal knowledge, and understanding of each other, on which mutual trust is built, allows the development of a realistic context for cooperation with the perspective of achieving a good science outcome. Mutual trust is necessary at all levels, not only at the top. Personal relations among scientists, as mentioned, are built during their research careers, such as reading papers published by others in the same discipline, meeting with each other during conferences, through professional discussions, and finally by previous cooperation experiences in data analysis, joint publications, and even joint mission proposals. However, this does not mean that they all know and trust each other even within the same research discipline. The scientific academic and research organizations as well as the space agencies can organize dedicated workshops or forums for them. In most cases, bilateral scientific workshops organized by the managing agencies planning cooperation programs or missions is a good first step in starting, or ice breaking, the connection for the scientific communities from both sides, providing opportunities to meet and gain mutual trust. This also includes personal relations promoted by joint events, like dinners, involving friendship and knowledge of the family environments.

From the point of view of cooperation at the agency level, the key element for success is to maintain full transparency for both specific projects and long-term plans. Publicly released policies and plans for science missions are a prerequisite for the mutual understanding of the science and programmatic objectives driving international cooperation. Moreover, program stability clearly fosters cooperation.

One usual way to gain transparency between the cooperative agencies is to have the management teams meet regularly, say once a year at minimum. These bilateral meetings, including personal relations built up at coffee breaks and joint dinners, are essential to strengthen personal relations within the management teams. In most cases, the information conveyed or underlining the formal speeches is more useful and even more important for making decisions. It is clear that a good personal relation among the managers may not necessarily lead to a cooperation mission, but it is certainly true that a bad personal relationship, lacking trust, will never end up with a successful cooperation agreement. In fact, you recognize your friends when you are down.

Finally, beyond the personal trust between scientists and agency managers, the political relationship among the governments involved in the cooperation is of vital importance. Governments, or their agencies, may consider the investment in science programs as part of their policy priorities, which could bring positive feedback from the cooperation. This is slightly different from the personal relations. In general, the personal relations between the scientists and the managers are not so much affected by high-level government policies. In fact, any scientific achievement should be considered a contribution to human civilization. However, it is a fact that the public funds used for space science missions are provided by governments, not by the scientists themselves. Political relations among governments are therefore essential at this point. Science should not be affected by what happened during the Cold War or even now, with new open conflicts. In fact, science should help improve relations, as a bridge between the partners, since they could probably only sign a science agreement and no others.

A risk not to be neglected is the possibility of the potential cooperative science mission not being selected during the competitive peer-review process. Stability is provided by a program structure with a regular sequence of missions of pre-defined "sizes." Regular calls provide sustained flight opportunities, a balanced program, and a perspective to the scientific communities involved. But they also allow the planning of national resources for scientific research and opportunities for international cooperation.

4.3 Leadership

This is another key element for successful cooperation. It is well-known that if nobody leads, nobody cares. Success requires good leadership, but there are many political and strategic elements to be considered in terms of the level at which the collaborating partners are leading. Cooperation is generally agency-to-agency, i.e., government-to-government, and private-sector collaborations are anticipated. The main aspect is who is taking the responsibility for critical decisions during the development phases of the mission, and who is ensuring a successful finalization of the joint project. In other words, a key element of cooperation is a clear model of ownership and governance, avoiding political chest-beating, where the role of stakeholders is maintained clearly. Experience shows that 50–50 agreements are difficult to manage, especially for large missions involving high technical challenges and budgets.

Top policy decisions involve the partners in specific parts. Cooperation, as earlier discussed, may involve the contribution of the launcher, the platform, ground support, or part of the payload. The issue of leading a cooperative mission generally stems from the incurred investment. The party putting in more funds should be the leading one. As we have discussed in the previous chapters, cooperation projects in space science are generally managed in a manner with non-exchange of funds. Each partner funds its respective contribution though the partners may not be equivalent in cost or effort. It is sometimes difficult to recognize who invest more than others, depending on local cost and the involvement of industry or public research organizations. It is generally accepted that the party leading the development of the spacecraft should lead the whole project. This is because on the one hand, spacecraft development involves more technical management and also faces more financial and schedule challenges. On the other hand, the launch vehicle is more on a ready product-level provision which does not need much new development or special management efforts. Other systems like TT&C and ground data receiving are less important in the mission, so they will not be considered as a vital contribution that could bring one party to the leading position.

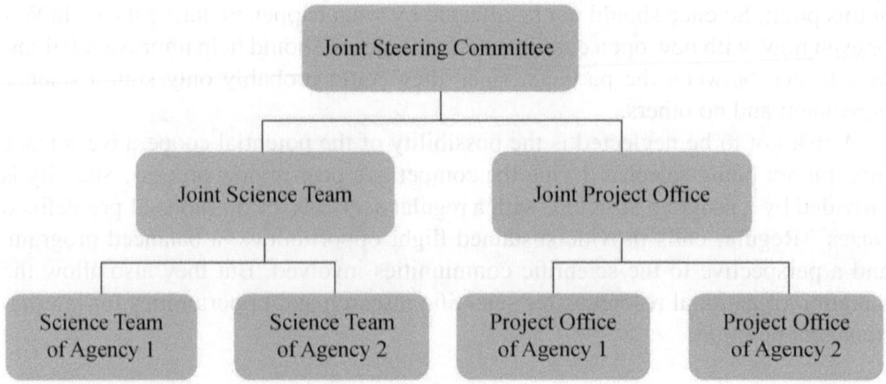

Fig. 4.2 The management structure of equal partners

Cooperation requires compromise. The party developing the spacecraft enjoys a leading position and must realize that they have the responsibility also to the partner who provides science payloads. The interfaces, including mechanical, thermal, and electrical ones, have to be negotiated with the cooperation partner from the design phase in order to meet the science objectives. A leading role does not mean that you can decide everything but also implies more responsibility for the overall science output.

Leadership, or who is leading the cooperative mission, must be written in the top-level cooperation agreement, such as the MoU which will be discussed later. It is an essential factor for a successful mission. The leading partner should shoulder risks and the junior partner is to be shielded by risks. Successful leadership is also proven by partner readiness to lead and follow in different cooperation opportunities.

However, in some cases, it is more complicated to define a leader in cooperation. In those cases, no matter who contributes what, the relation is equal and it is necessary to establish a joint committee. The structure of the management chart is shown in Fig. 4.2, where the members of a joint steering committee are from both agency decision makers. They meet annually and listen to the reports from the joint project office and the joint science team. However, the problems should be solved at a lower level first. The decisions that need to be made by the steering committee should have the green light already. Therefore, in this case, the joint project office is very important. They meet more regularly and all cooperation problems should be discussed and solved there.

4.4 Management

Implementing international cooperation projects clearly involves an increase in management complexity. Decision-making is inherently more complex with challenges in communication, different specifications, standards, and assumptions. In addition, interfaces are difficult to manage at a distance, with difficulties in monitoring progress and getting early warning of problems.

Cooperation must be based on clear agency-to-agency discussions and negotiations. Since they have the management and funding responsibilities, the structure of the agencies has the final word on proposed cooperative enterprises. Scientist-to-scientist discussions are of course welcome to shape the initial proposal, but they cannot replace the role of the agencies. In other words, scientists' wishes and decisions do not print money.

Joint projects must be designed and implemented within the scientific, technical, and budgetary capabilities of each partner. The management has to overcome difficulties to align schedules, budgets, and capability needs among the partners. Special attention has to be paid to "critical path" issues, minimizing the risk of over interdependence in critical areas. Clear interfaces involve the responsibility over full well-defined systems, while well-established agreements and monitoring of compliance are necessary in terms of funding, reviews, and tests. Otherwise, it would be impossible to manage any project to success. In order to carry out a science project in cooperation, its management structure has to be well understood and agreed on from the very beginning. Collaboration is structured to establish clearly defined managerial and technical interfaces to minimize complexity. Different development procedures have to be coordinated in an efficient way. Typical issues are the definition of the different development phases, including their naming, quality control standards, review procedures, or even the language used in the technical and management documents. A key issue to be tackled is the effect on the management plans of the differences in the funding cycles among the partners, as well as in the corresponding processes and decisions to be taken. Proper and timely funding must be agreed upon and respected to avoid cost overruns. With this objective, partners have to be reliable and avoid breaking mutual trust.

Since space technology was developed as classified due to its potential for defense, management procedures were developed in different countries independently. In particular, Russia and the United States followed very different schemes. Although Europe had a lot of cooperation with the United States, and also participated in some Russian science missions since the 1960s, the development process of space projects has its own identity. China received limited help from the Soviet Union in the late 1950s and the beginning of the 1960s, but it had a more independent space development, and the development procedures and quality control standards have distinctive characteristics. Similarly, India and Japan also have their own procedures. Differences also involve industrial policies, requirements, and costs. An important issue, to be mentioned later, is the access to test data of commercial subsystems and most relevant, the use of elements subject to export regulations.

However, no matter how the complete development phases are divided and called, they all fall into three key ones, namely the design, qualification, and flight model development phases, followed by launch and operations. In the design phase, the requirements are analyzed, and the spacecraft is designed to fulfill the science goals. In the qualification phase, the engineering model of the spacecraft is tested according to the approved quality standards, and all the issues found are solved to allow entering into production. The design parameters are frozen at the end of this phase. In the flight model development phase, or production phase, the final spacecraft that will be launched is manufactured and tested. At the end of this phase, a pre-shipment review will be carried out, and a green light may be given to ship the spacecraft to the launch site. The differences between different agencies may lie in the duration and detailed requirements of each phase. For example, at the design phase, one agency may require 18 months of development time to build a mechanical and electrical model, while the other agency may require a longer or shorter period or only an electrical model. In ESA, before the selection of a mission, there is a study phase while NASA uses the approach of formulation. A good tool to carry out these studies is the Concurrent Design Facility (CDF), where scientists and engineers work together to verify the proposed technical solutions to fulfill the science objectives. After the selection of the mission, there is a parallel design phase with at least 2 industrial contractors and a duration of 18–24 months finishing with a review leading to the mission adoption. A preliminary design review (PDR) is carried out, and the implementation phase starts with the definition of the mission and a selected industrial consortium. A critical design review (CDR) allows the project to enter into the full production phase. At the end, before the launch, a final acceptance review (FAR) is to be passed and, after the launch, a mission commissioning is carried out before starting the actual operation phase.

Internal and external reviews are essential in the development of any space science mission. They have to be clearly defined in scope and procedures and agreed upon by all the partners. In order to adjust and coordinate different development procedures from different agencies, the method of joint reviews is a necessary management tool, which intends to find a suitable time, independent of either partner development procedures, to organize a joint review to identify remaining issues that have not been covered by either procedure or the remaining interface problems. The most important joint review is then the joint pre-shipment review, since at this point, all developments from each side must be concluded and all problems have been solved already. The green light to ship the spacecraft to the launch site must be given by all parties participating in the mission.

Joint reviews are organized by the leading party of the spacecraft. In order to prepare a joint review, all documents, both managerial and technical, must be written in the language agreed in the MoU, generally English, and followed as jointly agreed. The review meetings, either online or presential, must also be conducted in the same agreed language. Besides the joint reviews, all other technical and management issues, such as scientific team meetings, joint tests, and joint onboard commissioning should all be managed in the same cooperative manner.

4.5 Cultural Differences

Among the problems that cooperation projects may find in their development, some are subtle and difficult to tackle technically. Cultural differences between the partners are an important aspect since cooperation is eventually between people. This is mainly visible in the case of Western and Oriental approaches to the development of their activities and social relations. The assumed ways of conduct may be totally different in these communities and lead to misinterpretations of messages and signals between partners.

The most interesting difference between the Western and Oriental cultures is how to say "no." In the Western culture, in most cases, to say "no" is not a problem, and the tendency is to say it as clearly as possible, in particular in the United States. However, in the Oriental culture, it is considered to be impolite to say "no" directly. They have the tendency to say "no" indirectly in order to avoid being impolite. In order to communicate in a more efficient way and try not to misunderstand each other, both sides should learn the cultural differences. A more obvious mistake that may occur is when someone moves his jaw left and right or up and down, you get the message of "no" or "yes."

However, all these misunderstandings can easily go away, when the partners get along well, become friends, and share social life. The cultural wall between them can be removed quickly when they sit at the table with some drinks. This brings another important cultural difference between different countries and regions. It is easier to make friends with Russians when you drink with them, but the drink should better be vodka. The same is with the French, if you know a bit about French wine and respect their choices of the wine on the table. For the Chinese, you are much respected, if you can master the chopsticks and even you can pick up peanuts using them. In other words, social life with your partners is the most important thing to avoid misunderstandings and overcome cultural barriers.

It is also important to remember the social benefits of joint space science programs for different countries and cultures. National space policy, identity, and vision, to maintain and enhance skills and capabilities, may drive their interest in space science missions but also foster the involvement of their society in the quest for knowledge and the fascination with space discoveries.

4.6 Legal Framework

Cooperation in space science is based on mutual trust, starting with mutual respect, but to ensure that the joint activities are carried out properly and timely, the understanding between the partners, the agreed responsibilities and work procedures, must be expressed in a written document, as already mentioned in previous sections. The written agreement generally takes the form of MoU, which is the most formal and highest-level agreement between two partners in space science cooperation. It should be signed by the responsible agencies/entities funding and managing the joint mission, such as between NASA and ESA, between ESA and JAXA, etc. Nevertheless, the MoU is not legally binding, and breaking it only affects mutual trust in future endeavors. In fact, legally binding documents are not welcome in cooperation partnerships based on mutual trust and good will to deliver.

In the MoU, the responsibilities and commitments of the institutions involved have to be described in a very clear way, such as who will provide what, who has the final responsibility of which system or subsystem, with the needed delivery times and actions to mitigate inter-dependency effects in case of delays. Since the MoU is a top-level management document, it is generally complemented with some detailed technical documents in attachments. In fact, under the umbrella of the MoU, more cooperation documents may be generated throughout the mission development phase, such as technical and management documents prepared for the joint reviews and other documents needed to define the relevant interfaces.

However, space science programs are driven by governmental agencies using public funds. Though the "no money exchange" principle is applied in this type of international project, any hardware development or provision of services involves some legal responsibilities that have to be clearly established. In this respect, it is important to consider that the implementation of the agreed contributions generally requires industrial procurements that are subject to international import/export regulations. Moreover, intellectual property rights emanated during the project development, either in industry or in public institutions, should be protected.

In case that one party is short of technology or some type of components, commercial contracts may also be involved in the cooperation. This will break the principle of no money exchange. Therefore, when a commercial contract is involved, the data sharing policy, if linked to actual contributions, should be modified and the contract be accounted to the buying partner. This may also be the case with any other exchange of hardware or services. Partners again should be aware of the existing legal framework for these activities to be successfully implemented.

Industry and public entities are subject to national laws with a significant impact on potential cooperation programs, like export restrictions for certain kinds of hardware or even technical capabilities. This is, in particular, the case in the United States through mechanisms such as ITAR, and in particular, the Wolf Amendment of 2011. In this section, we comment on these issues in order to inform possible cooperation partners in international space science projects. Scientists, and of course the signatories of an eventual MoU, should be well aware of them to design the joint activities in a pragmatic way and avoid possible frustrations.

Generally, international partnerships neither involve the joint development of technology nor involve products or processes that are potentially of near-term commercial value. Each party retains thus intellectual property rights in the technology/hardware that it brings to the partnership, which is developed independently of the other party. Clean interfaces are essential for these activities to be properly integrated from the very beginning of the cooperation. Some may require specific written documents, or legally binding agreements, to be produced depending on the national laws and regulations of the partners involved. Non-disclosure agreements may also be needed before the final agreement is reached. If the cooperative development yields an invention by the joint team, the intellectual property should be shared by both parties. This is generally described in the MoU and implemented by a joint patent application.

Of course, the results of the cooperation are fully shared, and the science is jointly exploited. The MoU includes the data sharing agreements, and scientists know how to handle intellectual property rights through their publications. Since the output of a space science mission is given by publications in peer-reviewed journals, agreements about the authorship, names, and order in joint publications generally solve the issue. All scientists in the space science field have the consensus that those who made contributions have to be included in the list of authors following the order of his or her contribution significance. However, the mission's name and supporting agencies have to be mentioned also in the paper or, at least, in the acknowledgement section.

Observational data are also considered to be part of the intellectual property rights. However, since all governments or space agencies want to reach the maximum science output, after a short period of exclusive use by the joint science team formed by the cooperation partners, the data will be normally released to the science community at large for anyone anywhere to be able to further exploit the scientific information provided. In this case, data users publish their results with the only request to acknowledge the mission's name.

Whenever the hardware or software developed by one partner in the cooperative relationship needs to be provided to the other, export restrictions must be considered. In most cases, since the cooperation is based on the principle of no money exchange, and the provision is viewed as a contribution to a joint cooperative science mission, it may not be considered a technology transfer. However, space activities go far beyond science and have implications that may lead to a regulation of the release of technologies to other countries that may threaten national security or commercial competition. This is the case of ITAR in the United States.

ITAR is a set of regulations developed by the United States administration to control the export and import of defense-related products and services. The purpose of ITAR regulations is to ensure that sensitive technology and equipment are shared with no control. Under the ITAR regulations, companies and individuals are required to obtain a license from the Department of State before engaging in the export or import of some products and services. Violations of ITAR regulations can result in civil and criminal penalties. The elements to which ITAR applies include common electronic devices such as infrared focal plane arrays and lasers that may be specifically modified or configured for military applications, as well as spacecraft (including research and commercial communications satellites and components).

Moreover, ITAR export rules apply to products developed anywhere using components or technologies of United States origin. This hampers the free flow of the global market and scientific cooperation in space, especially when the cooperation involves joint payloads. To partly circumvent the issue, European commercial satellite manufacturers have developed ITAR-free technologies to facilitate the export of their products and gain access to a larger service market.

A recent example of space science missions affected is the Solar wind Magnetosphere Ionosphere Link Explorer (SMILE), a joint mission between the Chinese Academy of Science (CAS) and the ESA with a planned launch in the first half of 2025. The SMILE mission is designed to study the interaction between the solar wind and the Earth's magnetosphere and the geospace dynamics. Even though it is a scientific mission, the division of hardware contributions had to consider ITAR regulations and the launch of the mission will take place from French Guiana.

More recently, restrictions on projects in collaboration with Russia were imposed by some European countries after the conflict in Ukraine, even for scientific activities. This situation affected well-developed cooperation plans for missions like Exobiology on Mars (ExoMars) to send a European robot to explore and drill the Martian surface or the scientific exploitation of the high-energy astronomy mission Spektr-RG with a key German payload element.

Chapter 5
International Organizations for Space Science

International organizations for space science have been created essentially to enable and foster international cooperation. Since the beginning of the space age, several international organizations in the area of space research have been created, such as the ad hoc COSPAR under ICSU, IAA, and IAF. Later on, when more and more space science missions were launched, new entities dedicated to scientific research in space were created like ISSI. Above all these, the United Nations has also created a dedicated office for space affairs called the United Nations Office for Outer Space Affairs (UNOOSA).

In this chapter, we will introduce some of those international organizations dedicated to space science or with an important connection to space science. We will concentrate on their functions and roles in international cooperation. In particular, we will examine their current activities and see how they can facilitate bilateral and multilateral international cooperation in space science.

5.1 COSPAR

COSPAR[1] is the most important international organization in space science. It was founded in October 1958, after the Soviet Union and the United States launched their first satellites, respectively. Its mission is to promote international cooperation in space research, to advance scientific knowledge about space, and to encourage the peaceful use of space for the benefit of humanity. COSPAR's primary goal is in fact "to promote on an international level scientific research in space, with emphasis on the exchange of results, information and opinions, and to provide a forum, open to all scientists, for the discussion of problems that may affect scientific space

[1] https://cosparhq.cnes.fr [2024-07-20].

© China Science Publishing & Media Ltd. (Science Press) 2025

J. Wu, A. Giménez, *International Cooperation in Space Science*, SpringerBriefs in Applied Sciences and Technology,
https://doi.org/10.1007/978-981-96-8366-6_5

research. This shall be achieved through the organization of scientific assemblies, publications or any other means."[2]

At the beginning, COSPAR had only 11 members. Some of them even with no space program at the time, such as Japan and India. However, they all understood that space research was a necessary tool to allow them to get into space. One of the most advanced countries then, the Soviet Union, joined the organization in the early 1960s.

The institutional members of COSPAR are national entities related to space research. However, each member country can only have one representative in the organization. This member institution, essentially space agencies, contributes an annual membership fee to COSPAR. Since there are not many countries with ongoing space research programs, even now, COSPAR only has 48 national members. Almost all countries that have space research programs are members of COSPAR.

The membership fee is divided into six categories. From the top level 1 to level 6, which represents the overall capabilities of the countries in space research. The biggest contributors, paying level 1 membership fees, are the United States, Russia, and China, since they all have independent capability to launch human programs into space. The second level reflects the capability of the member countries to manufacture and launch their own satellites. This kind of membership level evaluation goes all the way down to level 6, which represents the entrance level to space capabilities, such as some countries that have just established research groups in universities to use other member countries' data to do space research.

From the very beginning, COSPAR also accepted related scientific unions, within ICSU, to be its members. However, those members do not pay membership fees. The purpose of involving them as members is to allow an easy coordination of the space research activities between them and COSPAR, since space science is a multi-discipline domain and there are many overlaps. For example, there are cases where ground and space measurements have the same scientific objectives, such as the International Astronomical Union (IAU), the International Union of Geological Sciences (IUGS), and others.

The organizational structure of COSPAR is displayed in Fig. 5.1.

The council is the decision body of COSPAR, which is formed by the national and union representatives, the so-called members. The bureau is the executive body of the organization, led by the president and managed by the executive director. The key scientific body of the organization is the scientific commissions divided into eight disciplines as follows:

Commission A: Space Studies of the Earth's Surface, Meteorology, and Climate
Commission B: Space Studies of the Earth–Moon System, Planets, and Small Bodies of the Solar System
Commission C: Space Studies of the Upper Atmospheres of the Earth and Planets, Including Reference Atmosphere

[2] https://cosparhq.cnes.fr/about/charter/ [2025-02-06].

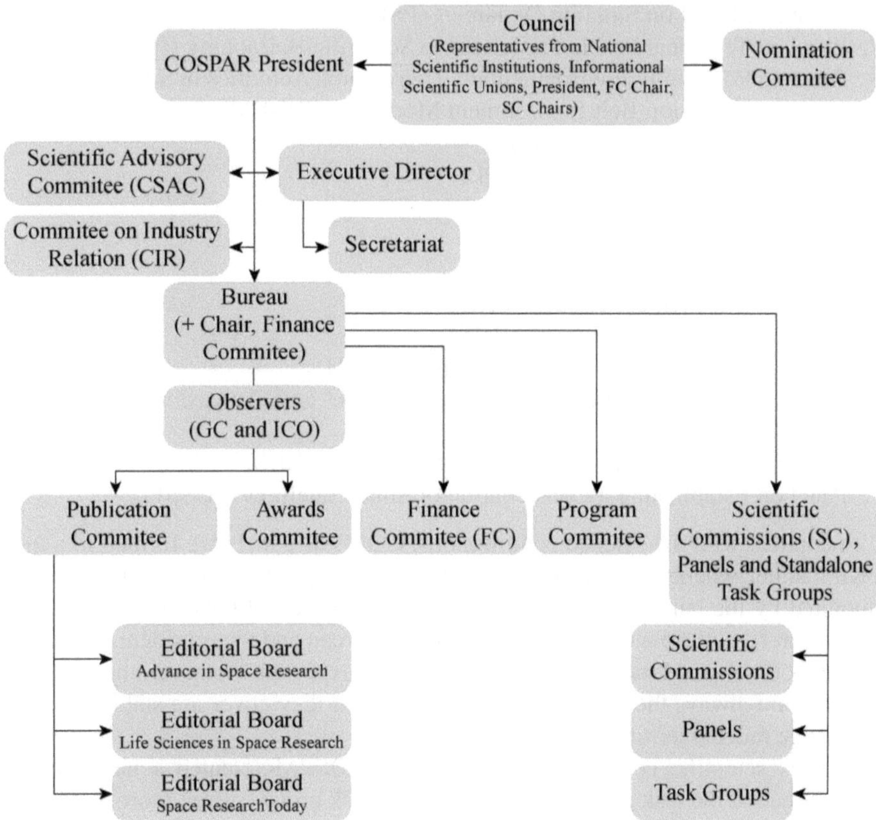

Fig. 5.1 Organizational structure of COSPAR

Commission D: Space Plasmas in the Solar System, Including Planetary Magnetospheres
Commission E: Research in Astrophysics from Space
Commission F: Life Sciences as Related to Space
Commission G: Material and Fluid Sciences in Space Conditions
Commission H: Fundamental Physics in Space

These commissions are formed on the basis of science disciplines, and the chairs of the commissions are nominated and elected in their own business meetings during the COSPAR scientific assembly by scientists of that particular discipline. Therefore, the COSPAR commissions are bottom-up structures reflecting the views of the international science community at large.

Since space studies deal with cross-disciplinary topics and technology-related research also needs to be covered, the COSPAR bureau has the right to define study panels in parallel to the scientific commissions. There are 15 panels as listed below.

1. Technical Panel on Satellite Dynamics (PSD)
2. Panel on Technical Problems Related to Scientific Ballooning (PSB)
3. Panel on Potentially Environmentally Detrimental Activities in Space (PEDAS)
4. Panel on Radiation Belt Environment Modeling (PRBEM)
5. Panel on Space Weather (PSW)
6. Panel on Planetary Protection (PPP)
7. Panel on Capacity Building (PCB)
8. Panel on Education (PE)
9. Panel on Exploration (PEX)
10. Panel on Interstellar Research (PIR)
11. Panel on Innovation Solutions (PoIS)
12. Panel on Social Sciences and Humanities (PSSH)
13. Panel on IDEA (Inclusion, Diversity, Equity, and Accessibility) (PIDEA) Initiative.
14. Panel on Machine Learning and Data Science (PMLDS)
15. Panel on Establishing a Constellation of Small Satellites (PCSS)

In contrast with the science commissions, COSPAR panels are formed in a top-down manner whenever they are considered necessary. The chair of each panel is appointed by the bureau and the president of COSPAR, not through internal elections. Therefore, the panels can be closed by the bureau and the president, whenever they think the job is done or no further advance is expected. However, the commissions remain always there, since the disciplines that they cover are stable and will not change much over time.

Besides scientific commissions and panels, COSPAR created a new type of structure called the task group. The mission of a task group is even more explicit in terms of the goals to be achieved and their short-term actions. Currently, COSPAR has created 6 task groups:

1. Task Group on Establishing a Constellation of Small Satellites (TGCSS)
2. Task Group on Establishing an International Geospace Systems Program (TGIGSP)
3. URSI/COSPAR Task Group on the International Reference Ionosphere (IRI)
4. COSPAR/URSI Task Group on Reference Atmospheres, including ISO WG4 (CIRA)
5. Task Group on Reference Atmospheres of Planets and Satellites (RAPS)
6. Task Group on the GEO (TG GEO)

The goals of these task groups are still to be achieved.

Due to the limited membership from countries that have space research activities, accordingly, the budget of COSPAR is also limited. However, in order to promote international cooperation, COSPAR has carried out substantial work as follows:

1. Organize a COSPAR general assembly every 2 years in even years, which covers all scientific disciplines of space research. Sessions are organized by all commissions, panels, and task groups. It usually has 2000 participants and is the largest scientific assembly of space research in the world.

2. Organize a topic symposium every 2 years in odd years, which covers a specific topic in space research proposed by the local organizer. Sessions are organized by the relevant commissions and panels together with the local organizer. It usually has a few hundred participants.
3. Organize roadmap studies and publish them as recommendations to space agencies. Studies are organized by the relevant commissions and panels, while the results are published in *Advances in Space Research*, a COSPAR journal, and will be sent to all major space agencies as input for their future program planning.
4. Promote and manage small (CubeSat) constellation programs under the no money exchange principle. This new initiative was just created in 2020 and is waiting for approval, to see if COSPAR can fill the gap where major space agencies do not have the interest to get involved.
5. Award recognition of scientists who have made significant contributions to international cooperation in space science. Since 2018, the awardee will have an asteroid named after him/her as agreed between COSPAR and IAU.
6. Organize space agency round tables at the general assembly to promote cooperation among them and provide updated information to scientists about potential cooperation opportunities.
7. Organize bilateral or multilateral meetings at any time necessary between agencies that may require COSPAR as a platform to talk to each other on potential cooperation activities.
8. Issue reports on the status of international cooperation. The first such report was issued in 2017.

5.2 IAA

IAA[3] is an organization very different from COSPAR. It is based on individual members, rather than institutional or agency/governmental members. Therefore, it is an independent non-governmental organization.

IAA was created by Theodore von Kármán in 1960 during the space race time. He served as the first president until 1963, when he passed away. At the time of its creation, he envisioned that the IAA should have its mission in four aspects.

1. Foster the development of astronautics for peaceful purposes.
2. Recognize individuals who have distinguished themselves in a branch of science or technology related to astronautics.
3. Provide a program through which the membership can contribute to international endeavors.
4. Promote international cooperation in the advancement of aerospace science.

[3] https://iaaspace.org [2024-7-20].

According to these goals, since the beginning, von Kármán invited the first group of noted scientists and engineers from the United States, the Soviet Union, and other parts of the world as the core group of members. Later on, the members are elected through very strict criteria, with nominations by IAA members only. Once a member is elected, the member pays a membership fee annually, ensuring independence from agencies or other public entities. The IAA's operation is fully dependent on the membership fees of individual members. The number of the current paying members, or the so-called active members, of IAA is around 1000. IAA also organizes symposiums and publishes books written by its members. However, the income from the symposiums and book publications is rather small.

IAA has a unique strength in getting well-known scientists, engineers, and astronauts together. Even if they do not speak the same language, IAA can always help them to communicate and sit together. The strong link to attract them together is that they are all recognized individuals.

The governing body of IAA is the Board of Trustees, which includes the president, four vice-presidents, four section chairs, four trustees from each section, the Legal Counsel, and the Secretary General.

The members of IAA are elected in four sections, which means that only the members within the same section can nominate others and be elected. The four sections are as follows.

Section 1: Basic Science.
Section 2: Engineering Science.
Section 3: Life Science.
Section 4: Social Science.

However, the studies organized by IAA are divided into six disciplines which are called commissions.

Commission 1: Space Physical Science.
Commission 2: Space Life Science.
Commission 3: Space Technology and System Development.
Commission 4: Space System Operations and Utilization.
Commission 5: Space Policies, Law, and Economics.
Commission 6: Space, Society, Culture and Education.

Apart from the discipline committees, IAA also has permanent committees as follows.

1. History Committee;
2. SETI Committee;
3. Small Satellites Committee;
4. Scientific-Legal Liaison Committee;
5. Space Debris Committee;
6. Space Traffic Management Committee;
7. Moon Farside Protection Committee.
8. Space Solar Power Committee

With these permanent committees, IAA keeps an eye on the important issues related to space. IAA also organizes many topic conferences and symposiums. Some of them are a series of meetings, such as:

1. IAA Planetary Defense Conference;
2. IAA Humans in Space Symposium;
3. IAA Symposium on Small Satellites for Earth Observation;
4. IAA Space Traffic Management Conference;
5. IAA Conference on Space Situational Awareness;
6. IAA IRG Interstellar Conference, etc.

IAA actually organizes more than 15 symposia a year and each of them has the number of participants ranging from 100 to a few hundred.

IAA also encourages international cooperation through its awards. An award called Laurels for Team Achievement is dedicated to significant contributions in international cooperation. Famous international cooperation missions in space science, such as SOHO, HST, Cassini-Huygens, Double Star, Cluster, and others, have all received this award.

IAA is a strong supporter of COSPAR and IAF and organizes the Academy Day at each COSPAR general assembly and International Astronautics Conference in order to let the members participating in such meetings get together.

5.3 IAF

Compared with COSPAR and IAA, IAF[4] is a more general organization covering all aspects of space activities. Its members are government agencies, public entities as well as industries.

IAF was founded in 1951, when scientists from the field of space research gathered in an attempt to ensure a constant dialogue between the space nations, regardless of any political turmoil. It was for sure that space was still a top secret mainly used for military purposes, such as the V2 rockets in Germany. However, after the World War II, cooperation was welcomed even in a sensitive field, such as space.

In September 1951, at the second International Astronautical Congress (IAC), the first constitution of IAF was signed by ten founding members in London. They were Argentina, Austria, France, Germany, Italy, Spain, Sweden, Switzerland, the United Kingdom, and the United States of America. After many years of development, IAF's missions are focused on the following six areas.

1. Promoting cooperation;
2. Advancing international development;
3. Sharing knowledge;
4. Recognizing achievements;

[4] https://www.iafastro.org [2024-07-20].

5. Preparing the workforce of tomorrow;
6. Raising awareness.

It is obvious that the organization is dedicated to promoting cooperation. However, along with the development of space technology, and after the creation of COSPAR in 1958, the activities of IAF concentrated more on space technology and cooperation in large-scale missions in general, such as manned spaceflight or deep space exploration. At its annual conferences, the so-called International Astronautical Congress (IAC), all space agency leaders meet regularly. Therefore, IAC is also considered an opportunity for policy exchanges for all space agencies around the world.

Different from COSPAR, IAF members mainly come from industries and there is no limitation to the number of members from a single country. Currently, IAF has around 500 paying members. Therefore, it is a big organization in terms of the number of members and budget.

COSPAR and IAF used to organize joint conferences every 10 years, called the World Space Congress, such as the 1992 conference in Washington, D.C. and the 2002 conference in Houston. However, the benefit of bringing scientists, space industry leaders, and engineers together was not very significant. The motivation was then compromised, and after 2010, the joint effort stopped. In order to continue it in smaller-scale meetings, the two organizations agreed to jointly organize a joint session at each IAC under the title of space science. Due to a relatively low attendance to the session, this effort is currently also facing problems. Another initiative, inside COSPAR now, tries to promote the cooperation by attracting more engineers to its general assembly instead of just using the platform provided by IAF.

5.4 ISSI

ISSI[5] is a research organization created by Johannes Geiss and a group of space scientists in 1995. It got support from ESA and the Swiss Space Office from its creation. It is located in Bern, Switzerland, and is dedicated to advancing the understanding of space science. ISSI tries to maximize the exploitation of space science missions in orbit by international teams of scientists, sharing knowledge about the best use of the existing tools. The institute provides a forum for scientists from different countries and disciplines to collaborate on space-related research projects, and it also hosts workshops, conferences, and other scientific meetings. ISSI's research covers a broad range of topics in space science, including solar physics, planetary science, astrophysics, and astrobiology, as well as Earth science from space.

ISSI is a non-profit organization registered in Switzerland as a science foundation. Currently, ESA, the Swiss Confederation, and the Swiss Academy of Sciences (SCNAT) provide the financial resources for ISSI's operation. The University of Bern contributes through a grant to one director and with in-kind facilities.

The organization structure of ISSI is shown in Fig. 5.2.

[5] https://www.issibern.ch [2024-07-20].

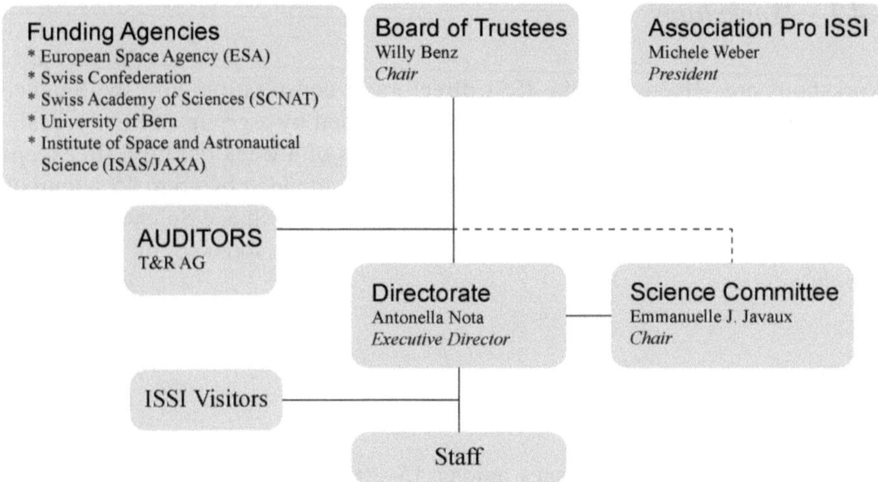

Fig. 5.2 ISSI's organization structure as of June 2023

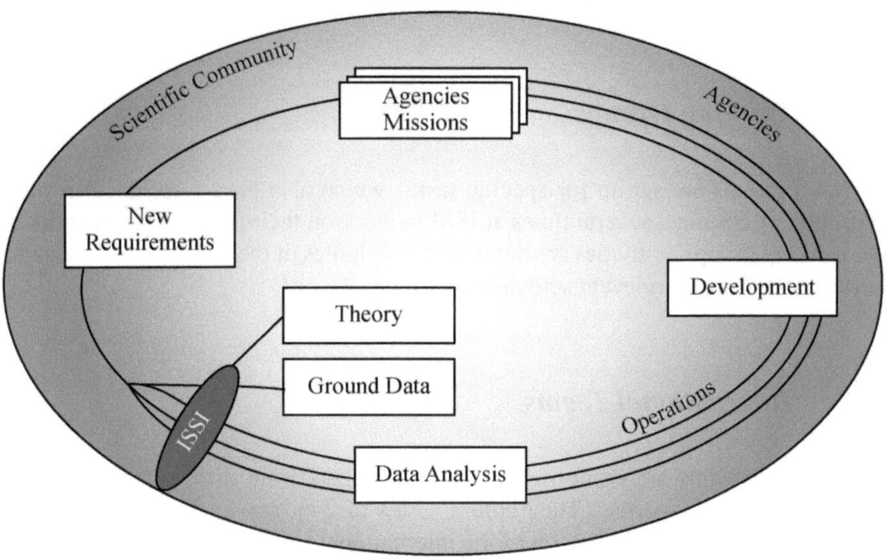

Fig. 5.3 ISSI and the data analysis by the science community

Different from other national research institutes, ISSI does not have many research staff. It uses its operational funds to invite and host scientists from all over the world to do studies at ISSI for short durations. When scientists arrive at ISSI in Bern, the lodging and per diem are paid by the organization. Therefore, it is a new type of institute that carries out studies by hosting visiting scientists, instead of recruiting them as staff, and letting them do the studies in house.

The functions of ISSI are shown in Fig. 5.3 and the tools used by ISSI to organize studies are in the following subsections.

5.4.1 Workshops

Workshops are selected by the ISSI directors in consultation with the Science Committee. The programs and speakers are defined by a group of highly qualified experts serving as conveners. The workshops of a week duration (in exceptional cases, a workshop can be repeated) may be attended by up to 45 scientists and experts. Workshops always lead to a volume of the Space Science Series of ISSI and, in parallel, issue papers in *Space Science Reviews* or *Surveys in Geophysics*.

5.4.2 Forums

Forums are informal free debates among 15–20 high-level participants on open questions of scientific nature or science policy matters. Forums do not necessarily lead to formal recommendations or decisions. They are generally held at ISSI for 2–3 days' duration (one time).

5.4.3 Working Groups

Working groups are set up for specific tasks, which also have a technical nature. Working groups meet several times at ISSI to work on their projects. The results of the working groups' activities are published as volumes of the ISSI Scientific Report Series (SR) or in the relevant scientific literature.

5.4.4 International Teams

International teams are composed of 8–15 scientists from different laboratories, nationalities, and expertise. They hold a series of 2–3 one-week meetings over a period of 18–24 months. The aim of the international teams is to carry out research projects leading to publications in scientific journals. The activities are chaired and organized by a team leader who is also the initiator of the proposal to ISSI. Though in close contact with the scientific staff of the institute, the international teams are largely autonomous in the execution of their projects. The international teams are set up in response to an annual call by ISSI. Proposals are evaluated and prioritized by the Science Committee.

5.4.5 *Visiting Scientists and Johannes Geiss Fellowship*

Although most of the studies conducted by ISSI are through the above tools which do not try to hold the scientists on a permanent basis at ISSI, it also invites a few visiting scientists, 1–2 per year and an honorary researcher called Johannes Geiss Fellow, 1 per year, to stay at ISSI for a longer time during the year when the fund is allocated.

With all these tools and research activities, ISSI staff and visitors publish around 200 research papers annually and several books as the results of the ISSI workshops.

5.5 ISSI-BJ

ISSI-BJ[6] is a similar international cooperation organization created by ISSI and NSSC in 2013. The purpose is to extend the model of ISSI to the East and in particular, to support the new space science programs of China and their internationalization. The Chinese Academy of Sciences provides the operational funds to ISSI-BJ, and NSSC provides all types of in-kind support to it, including the staff and operational costs. ISSI-BJ uses all the tools of ISSI and beyond them, ISSI-BJ also organizes forums to discuss the science objectives of new mission proposals which are not foreseen at ISSI in Bern.

5.6 UNOOSA

Since 1959, the United Nations (UN) has set up a committee called the Committee on the Peaceful Uses of Outer Space (COPUOS). Its permanent office is called the United Nations Office for Outer Space Affairs[7] (UNOOSA). COPUOS is reporting to the Fourth Committee of the General Assembly of the UN, and UNOOSA is a division of the office within the Department for Political Affairs.

COPUOS's responsibilities are, but not limited to, reviewing international cooperation in peaceful uses of outer space, studying space-related activities that could be undertaken by the UN, encouraging space research programs, and studying legal problems arising from the exploration of outer space. The committee has two subsidiary bodies: the Scientific and Technical Subcommittee and the Legal Subcommittee. The committee and its subcommittees hold sessions every year, respectively.

---

[6] http://www.issibj.ac.cn [2024-07-20].

[7] https://www.unoosa.org [2024-07-20].

The recent issues related to space science that have been discussed at COPUOS are:

1. Space weather;
2. Space debris and space traffic control;
3. Planetary protection;
4. Space resource utilization.

For space weather, COPUOS has established an initiative called International Space Weather Initiative (ISWI). It is a program of international cooperation to advance the space weather science by combining instrument deployment, analysis, and interpretation of space weather data from the deployed instruments in conjunction with space data and communicating the results to the public and students. It is not a cooperation program for space missions, but rather a program on instrumentation, data analysis, modeling, education, training, and public outreach.

For space debris and space traffic control, the UN has close coordination with the Inter-Agency Space Debris Coordination Committee (IADC), an inter-agency group of 13 members covering major space agencies, such as ESA. The UN adopted the IADC Space Debris Mitigation Guidelines issued in 2002 as the UN COPUOS Space Debris Mitigation Guidelines. It was endorsed at the UN General Assembly resolution 62/217 in 2007 and became reference guidelines to all members of the UN.

For planetary protection, COPUOS has overall responsibility but relies on COSPAR to provide the guidelines. According to the target you are aiming at, the protection strategy is divided into five categories from the highest, such as the Mars that may have both forward contamination and backward contamination, to the lowest, such as a small asteroid where there is no expected presence of life.

Space resource utilization is a more recent issue popping up with commercial space activities, after the United States, Japan, Luxembourg, etc. endorsed aggressive laws in order to encourage their firms to go after the available space resources. It shortly became a hot topic at the COPUOS meetings. However, there is still no consensus among the members of the UN to have a document to define what are space resources and whether it is legally correct to maintain them and sell them to others. This is certainly in contradiction with the 1967 UN space treaty. However, if the regulation is too strict, it may block the technology and even science of developing those resources for future human explorations and deep space adventures. A call to modify the 1967 UN Space Treaty has been raised due to this reason, but there is not a consensus so far.

Chapter 6
Examples of Space Science Collaboration Projects

As a complement to the previous chapters, we want to dedicate one chapter specifically to mention some cases of international space science collaboration. Of course, we cannot analyze all possible cases, as that would require several books to do so. Moreover, there are several books and reports available dealing with cooperative space science missions, such as the book by Bonnet and Manno (1994), the NRC-ESF study in 1998, and the Earth System Science Center (ESSC)-ESF's position paper in 2000. Nevertheless, most discussions focus on European cooperation with NASA and pay no attention to other space science actors. During the last two decades though, space science activities have flourished in other countries, creating new opportunities for collaborative endeavors.

In order to expand the study and demonstrate what has been presented in previous chapters, we will focus on a few typical examples of international cooperation in space science. These examples could serve as references for future international cooperation in science missions. As mentioned above, we have also decided to add some detailed explanations about missions in cooperation with China. The experience could certainly be extended to other countries and agencies.

6.1 The Halley Comet Encounter

Missions to encounter Halley Comet as it approached the Earth in 1986 were efficiently coordinated, showcasing an example of space science cooperation. A common goal of studying and characterizing Halley Comet was achieved by coordinating the flyby times and distances of six spacecraft, which were independently developed and launched. Information about the comet dust and gas environment models was shared before the encounter, as well as the scientific data retrieved afterward.

Halley Comet was chosen by the science community as a paradigm to study comets. The reasons included its brightness in its close encounter with the Earth, its

© China Science Publishing & Media Ltd. (Science Press) 2025

J. Wu, A. Giménez, *International Cooperation in Space Science*, SpringerBriefs in Applied Sciences and Technology,
https://doi.org/10.1007/978-981-96-8366-6_6

historic record of regular visits, and an excellent knowledge of its orbit. The broad scientific interest promoted international cooperation to fly the six spacecraft to observe the comet, which was known as the Halley Armada. The Soviet Union's Vega 1 and Vega 2 spacecraft[1] were launched on 15 and 21 December 1984, respectively. Japan's Sakigake[2] was launched on 8 January 1985 and Suisei on 19 August 1985. ESA's spacecraft Giotto[3] was launched on 2 July 1985 by an Ariane 1 launch vehicle from Kourou in French Guiana. Although these launch dates spread over a period of 8 months, all their encounters with Halley Comet occurred within a week in March 1986: 6 March for Vega 1 at 8890 km from the comet, 8 March for Suisei at 151,000 km, 9 March for Vega 2 at 8030 km, and 11 March for Sakigake at 7 million km. Giotto made a historic flyby of Halley Comet at a distance of about 596 km on 14 March 1986, the closest encounter by far. Finally, NASA's ICE mission could measure the comet from a distance of 28 million km on 25 March 1986. The images sent back by Giotto's Halley Multicolour Camera radically transformed the ideas about what the nucleus of a comet really looked like. Other components of the cooperation were certainly key to the success of the closest encounter.

Since the beginning of the mission planning, the space agencies involved were aware that the overall scientific return and mission success could be increased through cooperation. They agreed to form the Inter-Agency Consultative Group (IACG) in 1981 with the task of coordinating all matters related to their missions to Halley Comet and the observations to be carried out in space. Furthermore, an effort was made to complement the in situ observations with remote measurements from the ground which helped to put the spacecraft data into the proper scientific context. This was known as the International Halley Watch (IHW). This type of international cooperation to provide ground-based monitoring data was later extended to ESA's Rosetta mission to the comet Churyumov-Gerasimenko, as well as NASA's missions Deep Space 1 and Stardust. In addition, on 4 July 2005, ground-based telescopes around the world joined the Hubble Space Telescope (HST) and the Rosetta mission in monitoring the impact of NASA's Deep Impact spacecraft on the comet Tempel 1.

Cooperation on the Halley Comet encounter was a typical type of international cooperation among independent missions or programs from different countries or agencies. It could be called cooperation between programs or a cooperation program of programs. The leading body of the cooperation was the IACG which played a crucial role. This type of cooperation is featured by relatively independent mission management with the coordination of orbit design restricted by the nature of the comet return event, with no restriction from any of the participants. The benefits for all parties are in the science data sharing. Without cooperation, any single country or agency would have never obtained such rich data returned from many different

[1] https://link.springer.com/referenceworkentry/10.1007/978-3-642-27833-4_1650-5 [2024- 07-20].

[2] https://www.isas.jaxa.jp/en/missions/spacecraft/past/sakigake.html [2024-07-20].

[3] https://www.cosmos.esa.int/web/giotto/home [2024-07-20].

instruments, with different distances from the comet and different comet environments when the comet was approaching the Sun. For this reason, countries with different programmatic backgrounds, like the United States and the Soviet Union, could still work together. This sets up a model in which space science and exploration, as a topic for all humankind, offer joint efforts to respond to everyone's needs.

Following the success of the Giotto mission to Halley Comet, the science community worldwide realized the importance of additional visits to more comets, which could provide detailed analysis of these pristine elements to reveal the origin of our solar system. An ambitious comet-nucleus sample-return mission was proposed by ESA in collaboration with NASA. But, in the mid-1990s, it turned out that NASA would no longer participate. The project became an ESA-led mission, which was redefined as a cometary orbiter with a lander. The mission, called Rosetta, was launched on 2 March 2004, and stands as the most significant successful contribution of European space science to our understanding of comets.

6.2 The Hubble Space Telescope

The Hubble Space Telescope (HST)[4] (Fig. 6.1) was an ambitious project that required a combination of different cooperation schemes. With the strong leadership of NASA, ESA and the Canadian Space Agency (CSA) joined in the project to have the most powerful astrophysical tool in space. The proposed observatory was placed in low-Earth orbit, using the retrievable space shuttle and a Canadian robotic arm to deploy instruments during its operations. At some 590 km altitude in an orbit of

Fig. 6.1 HST

[4] https://science.nasa.gov/mission/hubble/ [2024-07-20].

around 96 min, the telescope can be accessible to shuttle flights, which allowed the first space telescope to be serviced by astronauts, including the change of instruments, repairing subsystem failures, or lifting the orbit. In fact, the optical polishing defect found in the main mirror was solved with the COSTAR optical correction system which was installed during one of the first several servicing missions. HST was launched on 24 April 1990 with Space Shuttle Discovery (STS-31) and is still in operation after more than three decades.

The project was led by NASA as the first of a series of large space observatories to explore the universe at different wavelength ranges and unprecedented capabilities, mainly in resolution and sensitivity. The scientific goals of the mission were, of course, of high interest to the world community, and ESA joined at the beginning of the design and development of the mission. The European contribution was not only limited to the science part but also to the spacecraft, providing the system for the deployment of the solar panels and a focal plane instrument, as well as the operation and data processing of the mission. Canada provided the robotic arm for the deployment and further servicing operations.

Science operations required the establishment of the Space Telescope Science Institute (STScI) in Baltimore to help the science data retrieval and processing. Furthermore, the STScI also provided support to the scientists in the application of observing time and the use of the telescope. A European element with a full copy of the data archives was installed in Garching (Germany) at the ESO offices. After more than a decade, with the increase in Internet connection and facilities, the European contribution was no longer necessary. According to the cooperation agreement between ESA and NASA for the operation of HST, Europe contributes 15 ESA staff to the activities at the STScI in Baltimore. International cooperation was clear in terms of hardware and operations, but the new approach was about access to the observatory.

HST is operated as an open observatory. All scientists across the world can apply for observing time which is allocated after a competitive review of the proposals by a single allocation committee. There is an initial guaranteed time for the groups providing instruments and open time available to all scientists. There is no separate allocation committee for observing time corresponding to the contributing partners, i.e., NASA, ESA, and CSA. Nevertheless, the NASA-ESA agreement establishes a check of the successful proposals led by European scientists, after a number of time allocation periods awarded, with the aim of achieving at least 15% of the allocated observing time. If this was not the case, active measures to recover the agreed percentage should be taken. Nevertheless, this has not been ever necessary throughout all the years of operation of HST, indicating the competitiveness of the European science community as well as the mixed situation of both communities, working actively together.

HST is a type of international cooperation very different from the Halley Comet encounter. It was a kind of cooperation down to the system level, i.e., one party providing instruments or subsystems on board another party's spacecraft. In this case, the leading party, NASA, must be the one who has full responsibility for the spacecraft. The nature of the cooperation requires full trust between the partners, not only among the scientists that eventually compete for the observing time, but between the engineering teams and their agencies, based on a continuous and fully transparent exchange of information. Before the cooperation starts, an agreement should be signed between the parties. All important matters related to the success of the cooperation, such as responsibility, detailed engineering interfaces, and data sharing must be defined in the agreement. The success of the cooperation scheme of HST was later adopted by the new James Webb Space Telescope (JWST) which is currently in the operational phase.

6.3 Cluster and Double Star

Cluster[5] and Double Star[6] are cooperative projects between ESA and China implemented during the last three decades. The geospace was previously understood as a stable standard model with a magnetopause on the day side of the Earth at about 10 Earth radius (Re), a long magnetic tail on the night side, and two cusp regions over both North and South Poles. However, from the 1980s, more and more scientists believed this model should be updated. It should be a dynamic one changing together with solar wind input. For a dynamic magnetosphere, it is, therefore, impossible to measure with a single satellite since one cannot determine whether the changes measured are due to the crossing of boundaries or due to the time variation of the overall structure.

In the middle of the 1980s, a multi-spacecraft mission called Phoenix (later changed to Cluster) was then proposed (Escoubet et al. 2001). The mission was composed of four spacecraft flying in a large polar orbit with an apogee of 20 Re and a perigee of 4 Re. It was approved by ESA and the planned launch time was mid-1996.

Since the orbit of the Cluster mission is polar, there was a lack of information from the equatorial plane. The Cluster team was then looking for joint measurement partners. A German small satellite mission called EQUATOR-S joined the campaign (Haerendel et al. 1999). The launch date was set just 1 month after Cluster. It is known that the launch of Cluster by Ariane V failed. EQUATOR-S mission went into space successfully. However, it was damaged by a serious solar storm after 5-month operation. When ESA decided to resume the Cluster mission in 1997, there were then no corresponding missions to cover the magnetosphere in the equator region.

In order to get the maximum science output, at the beginning of the Cluster mission, ESA decided to open the data to scientists all over the world. Back then, Liu Zhenxing, a Chinese professor, had an extraordinary proposal and was thereby entrusted to build the Chinese Cluster Data Center at the Center for Space Science

[5] https://sci.esa.int/web/cluster [2024-07-20].

[6] https://english.nssc.cac.cn/missions/PM/201306/t20130605_102885.html [2024-07-20].

and Applied Research (CSSAR) in Beijing (later renamed as NSSC). At the beginning of 1997, Liu's team proposed a Chinese mission called the Double Star Program (DSP). The mission objectives of DSP were to investigate the physical processes of magnetospheric space storms, establish physical models of magnetospheric space storms, and develop dynamic models and prediction techniques of the near-Earth space environment (Liu et al. 2005). It has two small satellites flying over both the polar region and the equatorial region. Cluster and DSP later became joint China-ESA cooperation. The second launch of the Cluster mission was successfully conducted in mid-2000. Therefore, many of the flight model backups can be used on DSP. In this case, China is leading DSP and is responsible for the DSP's spacecraft design and implementation. ESA is responsible only for the science payloads provided to DSP. At a higher level, the science data acquired by Cluster and DSP are shared between the European and Chinese science communities.

The two spacecraft of DSP (TC-1 and TC-2) were launched in 2003 and mid-2004, respectively. They had been operating in space for about 7 years and ended in 2010. In the same year, IAA awarded the DSP/Cluster team the IAA Laurels for Team Achievement Award. Through the six-point joint measurement, DSP has yielded great scientific output, e.g., the discovery of density holes in the solar wind from magnetopause and the discovery of large-scale magneto-tail oscillations. Up to 2010, DSP had already produced more than 200 publications, including those published in the *Journal of Geophysical Research*, *Annales Geophysicae*, *Geophysical Research Letters*, *Chinese Journal of Geophysics*, *Chinese Science Bulletin*, *Chinese Journal of Space Science*, etc.

First, looking back to the implementation of the cooperation between Cluster and DSP, we can say that it was cooperation between two programs or missions. Therefore, it had clear interfaces and was easy to implement. It was a typical "1+1>2" cooperation model. Second, it was also cooperation between systems, i.e., one party integrated its hardware into the other's. It is a more complicated type of cooperation, which needs more mutual respect and trust between the partners. However, Cluster and DSP performed well in these aspects. The harmonious relationship was observed at all levels, from scientists and engineers to management. However, the most important is the science return. As explained above, multiple points' measurement of the dynamic magnetosphere is so much demanded that no one wants to lose the opportunity to add more measurement points to their own missions. In this sense, the cooperation initiative is bottom-up rather than top-down. But once the scientific benefits are there, the top-down management decision is certainly key and assures the smooth development during the entire cooperation period.

6.4 INTEGRAL

The science community worldwide was very much interested in exploring the high-energy domain of the radiation that comes from explosive sources in the universe, after several successful exploratory missions, such as the European COS-B

Fig. 6.2 INTEGRAL

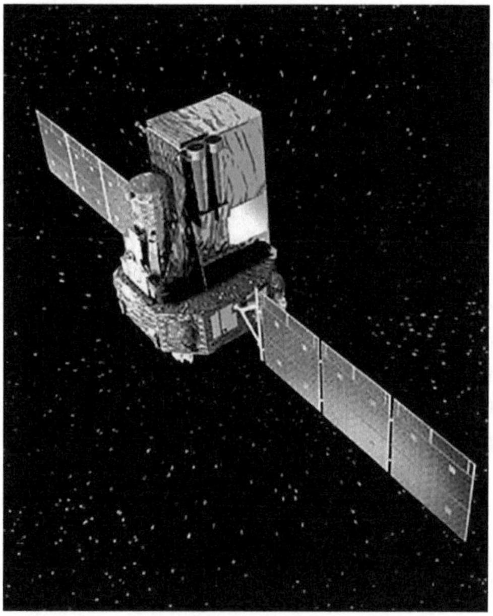

launched in 1975 or the Soviet Union's Granat launched in 1989 with a strong contribution of French scientists. In fact, in 1989, a group of scientists led by Anthony J. Dean (University of Southampton) and Jim L. Matteson (University of California, San Diego) proposed to ESA the mission INTEGRAL[7] which was selected as the next ESA medium-size scientific mission (M2) of the "Horizon 2000" program (Fig. 6.2).

The design of a high-performance gamma-ray observatory was further developed in Europe by a group of French, Italian, and German scientists together with the United States' high-energy scientists, who counted with the late Neil Gehrels in the science working team of the mission. Unfortunately, the initial plans to have another chance of ESA-NASA cooperation, such as HST, with Europe taking the leading role this time, were not fructified. NASA withdrew from the provision of a spectroscopic instrument, keeping only the ground station support of the Deep Space Network. NASA launched the Compton Gamma-Ray Observatory in 1991 and the explorer HETE 2 in 2000. The payload gap was filled with a consortium between French and German astronomers, led by Gilbert Vedrenne and Volker Schönfelder, while the problem with the launcher had to be solved by means of an agreement with Russia. A Proton launcher was made available, and the Russian scientists, with Rashid Sunyaev in the science working team, obtained a certain amount of guaranteed observing time to be allocated by their own time allocation committee. The main part of the observing time was allocated through an open competitive proposal process from scientists around the world.

[7] https://www.esa.int/Science_Exploration/Space_Science/Integral_overview [2024-07-20].

INTEGRAL was eventually launched by a Proton rocket from Baikonur on 17 October 2002 and was placed in a highly elliptical orbit with a 72-h period. Shortly after that, NASA launched the Swift mission for gamma-ray bursts (GRBs) in 2004 and Fermi (initially GLAST) in 2008. Like INTEGRAL, scientific collaboration in data sharing and open access to observing time was fully implemented.

INTEGRAL is a cooperation type with very clear interfaces where one party provides the spacecraft and the other provides the launcher. Both parties get observing time in proportion to their contributions to the whole mission mutually agreed upon through negotiation and recorded in the written agreement. This type of cooperation either happens at a late stage of development or between parties that do not wish to be involved in complicated managerial issues. It can be considered as a model for cooperation between countries or agencies where complete trust has not been established, or as the starting point for new cooperation partners.

6.5 Cassini-Huygens

The Cassini mission to explore the Saturn rings and its moons was a NASA-leading big project that made use of Radiation Thermoelectric Generators (RTGs) to provide energy to the spacecraft at large distances from the Sun. European scientists soon identified the mission as an opportunity to have access to the outer solar system, and ESA agreed to provide the Huygens probe to land on the Titan as a contribution to the joint Cassini-Huygens[8] mission to the Saturn. Essentially, NASA wanted to provide a ride to the ESA Titan probe in exchange for the joint science exploitation of all the data obtained by the instruments on board both elements. In addition, ASI provided the high-gain antenna of the Cassini spacecraft (Fig. 6.3).

ESA did not have the technical capability to fly beyond the Jupiter due to the performance restrictions of solar panels at longer distances. Cooperation with NASA provided access to the Saturn with the batteries of the Huygens probe fully charged before decoupling from the host spacecraft Cassini. In addition, the host spacecraft also provided communication and data links for the lander. It should be mentioned that during the long mission flight to the Saturn, a potential issue with the communication link between Cassini and Huygens was identified as a result of the expected frequency changes due to the Doppler effect in the signal during the approach and landing on the Titan. The ESA team at the European Space Operations Centre (ESOC) collaborated with NASA specialists to find the solution, which was achieved by modifying the flight dynamics of the encounter of Cassini with the Titan and the release of Huygens.

[8] https://science.nasa.gov/mission/cassini/ [2024-07-20].

Fig. 6.3 Cassini-Huygens

Cassini, with Huygens, was launched on 15 October 1997 from Cape Canaveral by a Titan/Centaur rocket. After the mission entered into orbit around the Saturn almost 7 years later, Huygens was released from Cassini on 25 December 2004, on the Christmas day. On 14 January 2005, Huygens reached the Titan and descended into its dense atmosphere and finally landed on the surface. Spectacular images and scientific data were obtained in this historical landing of a human artifact in a new world, the farthest-away landing ever.

This collaboration scheme was later used by ESA and JAXA in the mission BepiColombo to study the planet Mercury, which was launched on 20 October 2018 from Kourou in French Guiana by the Ariane-5 rocket. This time ESA is the leading agency, hosting a magnetospheric orbiter provided by JAXA.

This is a typical type of cooperation in deep space, where one agency leads the overall program and the other takes the ride. However, for the Cassini-Huygens mission, it is worth mentioning that when the communication issue was detected, the mission was already on the way to the Saturn. Without close cooperation between NASA and ESA, the problem would never have a chance to be solved, and the Huygens landing would become a dead mission. Instead, after a recalculation to replan the trajectory before the release of the Huygens spacecraft to the Titan, the Doppler effect was eliminated to the minimum and was within the bandwidth of the receiver. The lesson learnt was that giving a ride to your partners is not as simple as just providing more payload resources. The cooperation should last for the whole mission, from beginning to end.

6.6 SVOM

The Space-based Multi-band Astronomical Variable Objects Monitor (SVOM) mission[9] is a Sino-French cooperation project in space science. The initial idea came from a French proposal of a wide field-of-view high-energy instrument called ECLAIRs and a proposal from China to observe GRBs. Discussions to form a joint mission started as early as 2004. At first, the contributions from both sides were almost equal. For example, CNES would provide the spacecraft platform and half of the instruments, and the China National Space Administration (CNSA) would provide the launcher and the other half of the instruments. However, along with the negotiations, ITAR issues were the main obstacle. In the end, it turned out that CNSA provided both the spacecraft platform and the launcher. CNES provided half of the instruments and also the ground data receiving station network.

The cooperation management format was decided from the very beginning. The equal partner management structure remained with a joint steering committee on the top and supported by the joint science team and the joint project office.

The science objectives of SVOM are the following:

1. Detect and quickly locate various types of GRBs;
2. Measure the electromagnetic radiation properties of GRBs;
3. Study the evolution of the universe and dark energy by observing GRBs;
4. Make fast follow-up observation of astronomical transient sources such as those linked to gravitational waves.

For the science instruments, the Chinese side is responsible for the Visible Telescope (VT), the Gamma-Ray Monitor (GRM), the Ground-based Follow-up Telescope (GFT), and the Ground-based Wide Angle Cameras (GWAC); the French side is responsible for the low-energy gamma-ray imager and spectrometer ECLAIRs, the low-energy X-ray telescope LE, and the ground VHF network receiving system GFT.

Once the mission configurations and contributions from each side were decided, the cooperation entered into a full-speed development phase. Unfortunately, the pandemic slowed it down for another 3 years. SVOM was successfully launched on 22 June 2024 and it is now in full operation (Fig. 6.4).

[9] https://www.svom.eu/en/home/ [2024-07-20].

Fig. 6.4 SVOM

6.7 SMILE

After the successful cooperation on Cluster and Double Star, ESA and NSSC began discussing the potential for future cooperation in space science. The approach was to hold a bilateral meeting between the two agencies every year. The meeting is held annually, with the location alternating between Europe and China. This mechanism started in 2004 and continued until it was interrupted by COVID-19.

Solar wind Magnetosphere Ionosphere Link Explorer (SMILE)[10] is a mission fully generated by this bilateral meeting mechanism. At the meeting in 2015, both parties reached the consensus that if a new cooperation mission should be generated from scratch, it must follow the bottom-up principle. Therefore, two workshops were organized by the cooperating agencies. At the first workshop, there was no condition imposed on the science community; anyone could propose any mission even without a partner from the other side. There were more than 40 new mission proposals presented and most of them were without cooperation partners. And during the workshop, some of those proposals received special attention and triggered discussions between scientists from both Europe and China. This laid down a solid foundation for the second workshop. During this second workshop, the number of proposals shrank to 25, as a result of the agencies' requirement that all of them should be presented by joint teams between Europe and China, and SMILE was one of them (Fig. 6.5).

According to the agreement between ESA and NSSC, the next step was to issue an open call to invite joint mission proposals. After this call, 13 proposals were received. Immediately after that, ESA and NSSC carried out a technical screening study independently at their respective Concurrent Design Facility (CDF). During

[10] https://english.nssc.cas.cn/smile/overview/ [2024-07-20].

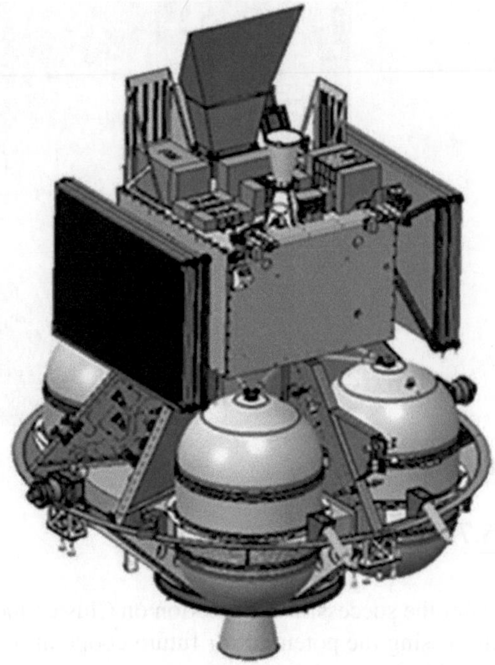

Fig. 6.5 SMILE

the studies, the joint teams were invited to present their proposals again at the CDFs. The lists of selected proposals passed the screening on both sides, and the ranking of proposals was then shared. It turned out that both sides soon agreed on which were the most promising proposals.

The last procedure of down-selection was based on the scientific impact of individual proposals. The scientific evaluation was made during a competitive process with an independent panel of European and Chinese scientists. The joint committee finally proposed to ESA and NSSC to jointly fly SMILE to study the dynamic interaction between the solar wind and the magnetosphere of the Earth, using soft X-rays and ultraviolet cameras to monitor the magnetopause, the polar cusps, and the auroral oval. In addition, the simultaneous in situ measurements will be taken with an ion analyzer and a magnetometer. SMILE is expected to be launched in 2025.

Chapter 7
The Future of International Cooperation in Space Science

In recent years, we have witnessed a global change in the planning and development of science missions. New actors are flourishing, mainly in Asia, but also in Latin America and Australia. In addition, lower-cost access to space and the introduction of new technologies, such as miniaturized systems with improved performance, open new ways of cooperation and require changes in traditional methods of carrying out collaborative science projects. The basic underlying mechanisms described in the previous chapters will certainly remain, including the science goals, mutual benefit, mutual trust, and cultural understanding.

7.1 Important Science Frontiers Need Joint Efforts

The frontiers that space science would address are among the most important domains of all science disciplines, such as the search for extraterrestrial life and the development of Earth life in space, the evolution of the universe including dark matter and dark energy, the prediction of solar activity and its effects on the Earth and human beings, the global climatic change and its effects on the Earth's biosphere, etc.

All these important science goals are international in nature, i.e., they are fundamental issues and questions for all human beings. Therefore, solving or answering them needs the joint efforts of scientists across the world. In fact, the answers and solutions belong to all human beings.

It is true that the funds for space science programs are mainly governmental in all space countries with space science missions. However, all these governments have the responsibility to make contributions to science for all human beings, not to mention that most of them are big countries like the United States, Russia, or China, and economic powers like Japan and major countries of Europe, who should shoulder global responsibilities toward mankind.

© China Science Publishing & Media Ltd. (Science Press) 2025

J. Wu, A. Giménez, *International Cooperation in Space Science*, SpringerBriefs in Applied Sciences and Technology,
https://doi.org/10.1007/978-981-96-8366-6_7

Besides, as it is said, if you want to go fast, you go alone; if you want to go far, you go together. For a particular country, it is also true that if it wants to go fast, it can implement a space science mission without international cooperation, but the impact may become smaller, and the cost may be higher. If it chooses to implement it with international cooperation, as discussed in all previous chapters, it may have a chance to produce much better science output and lay down the foundation for future cooperation, yielding a much larger impact. In this chapter, we will look at the new factors that will influence the future international cooperation in space science.

7.2 Commercial Space as a New Player

Since the 1970s, more and more commercial space users emerged such as providers of global TV broadcasting, long-distance or overseas telecommunications, private remote-sensing services required by meteorological forecast, precise agriculture, and education. Compared with governmental space endeavors, these providers demanded much lower costs of launch services and satellite manufacture.

Since the 1980s, compared with the previous government/agency-driven programs, satellite manufacturing has become more and more open. Universities and small commercial companies were new players using off-the-shelf commercial electronic components which significantly reduced the cost of satellites. The mass of satellites could be reduced from tons to hundreds of kilograms or even several kilograms. Satellites were then divided into the following categories in terms of weight, i.e., large satellites (>1000 kg), small satellites (>100 kg), microsatellites (<100 kg), or even nanosatellites (<10 kg), CubeSats (~1 kg), and pico-satellites (chip-satellite, much less than 1 kg).

Since the 1990s, more and more universities have been using nanosatellites for educational purposes. A standard design was created called cubic satellites. The standard size was one cubic decimeter, as shown in Fig. 7.1. All components used in this kind of design quickly became commercial, available off the shelf. If cubic satellites cannot fulfill the design requirements, we can use two or three, or even

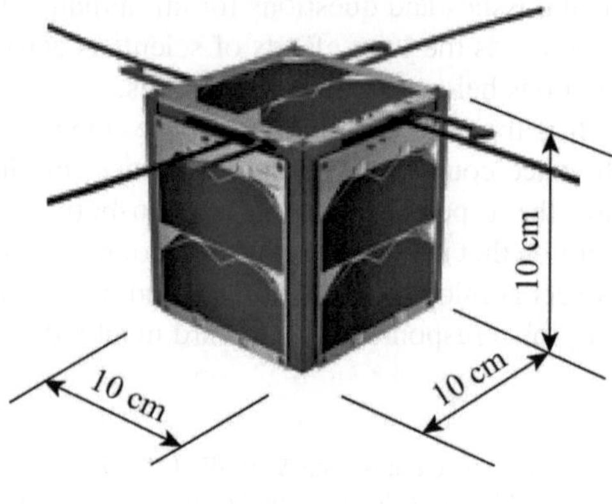

Fig. 7.1 A typical cubic satellite or CubeSat

more cubic decimeters and connect them together, called 2 units or 3 units CubeSats, or 2U, 3U up to 16U. Generally speaking, if the manpower costs are not counted, the CubeSat program's costs of a university would range from 10,000 USD to a few hundred thousand USD.

The satellite itself is not the only cost of a space mission. The cost of launch service should also be considered. Among other solid surface celestial bodies in our solar system, the Earth's gravity is the largest. It means that it is more difficult to leave the Earth. According to Konstantin E. Tsiolkovsky, a Russian space scientist, in any attempt to increase your speed when you want to leave the Earth, the power you need increases exponentially. Even if you want to send up only a few kilograms of payload, the threshold is high. The effort to reduce the launch cost started with Elon Musk in 2006. He introduced the multi-rocket-engine design to reduce the cost. In 2018, he finally succeeded with the reuse of the first stage of the rocket, which is the most expensive part of the launcher. At the beginning of the twenty-first century, the cost to send 1 kg payload into low-Earth orbit (LEO) was around 50,000 USD. It is now reduced to around 1500 USD with a Falcon 9 launcher provided by SpaceX.

7.3 Constellations of Small Satellites for Space Science

Whether a big satellite can be replaced by an array of several small satellites while keeping the same performance has been an issue for discussion in the past years. In fact, this depends primarily on the payloads of the mission. If the main payload is a large optical telescope, a combination of several small telescopes will, in fact, introduce more difficulties. Several small satellites with small telescopes on them need several satellite service systems. In particular, the attitude control systems must be coordinated to point the several small telescopes precisely. In addition, the distances between each of those small satellites should also be measured precisely in order to retrieve the composed images. This array optical method is called the interferometric imaging method. However, whether it is a cost-effective design is still unclear. Thus, the constellation of small satellites for optical imaging is still not ready for implementation.

Similar to the optical method, if we work in the microwave or even lower-frequency bands, the interferometric imaging with a constellation of small satellites is much more feasible. This is because, in the microwave band, it is difficult to build and launch a single antenna with an aperture as large as dozens of meters or even larger, which means the angular resolution of a single satellite will never be as good enough as that in the optical band. Even for astronomical observations, the aperture of a microwave antenna needs to be as large as hundreds of meters. For a lower-frequency band, say 30–150 MHz band, the aperture needs to be in the order of kilometers to have useful science return. Therefore, using an array of small satellites for microwave imaging becomes an attractive way to do space science research. There are several such missions currently under development. It is certain that in the

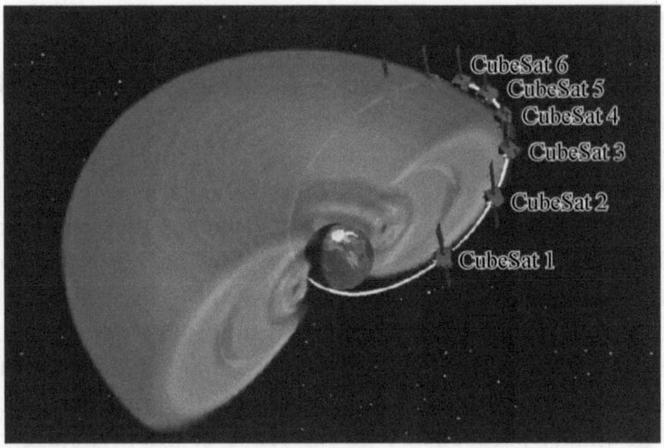

Fig. 7.2 CORBES: A small satellite constellation to measure the outer radiation belt

microwave or lower-frequency bands, a constellation of small satellites has better performance than that of a big single satellite. However, due to the quite unique technical requirements of those small satellites, very strict calibrations between them are needed. If international cooperation is considered for such a mission, it requires the partners to have close relations, trust, and technical transparency.

However, there are cases where the constellation of small satellites does not need a single or identical design and close coordination, such as the mission to explore the Earth's magnetosphere. A single satellite will never get enough information about it, due to its changing structure with time whenever a solar event is triggered. However, a cluster of satellites flying together at a distance from a few hundred to a few thousand kilometers can identify whether the changes come from crossing the boundary of the magnetosphere or from transit events. Every satellite could do its measurement alone, and if we put the data together afterward, we can then obtain the overall information about the magnetosphere. The requirements for such a constellation include position measurement with moderate accuracy and cross-calibration of sensors, such as magnetometer and particle detector, which can be done before launch or even in space, when they are passing through the same stable environment. Compared with the interferometric imaging arrays, this cluster type of constellation is much easier for international cooperation. Figure 7.2 is one such example called Constellation of Radiation Belt Survey (CORBES) (Wu et al. 2024) proposed and coordinated under the small satellites for space science initiative (Millan et al. 2019) of COSPAR.

7.4 Commercial Services for Space Science

As a traditional government-dominated area, space science has suffered from limited funding since the beginning. With the fast advancement of the new space sector, the cost of a space mission will also become lower and lower if governments allow the mission management to use the services provided by commercial space companies.

Apart from the direct procurement of satellites and launch services, institutions that are carrying out research can now also get services for space science missions from commercial space companies, such as receiving particular images from commercial remote-sensing satellite companies for Earth science study, flying payloads of microwave gravity for fluid-physics and life science experiments, getting observation time from commercial space telescopes, and receiving data from commercial ground stations or relay satellites. Those commercial companies may come from any country around the world. In case there is a no money exchange agreement signed, it is also considered as international cooperation since the country that pays for the science data may find one way or another to repay the commercial companies for their services.

For scientists, in very special cases, they could seek support from commercial sectors or get funds from private foundations. For support from commercial sectors, they may ask for advertisement and visibility through the mission. For private foundations, they may have a special interest in certain subjects, such as extraterrestrial life search, global change, and the prediction of earthquakes.

7.5 Cooperation and Competition

In many cases, cooperation is not the only means or possibility to promote science. In space programs, competition exists in different phases even within the design and development of a science mission. In general, competition and cooperation are mixed: cooperation in sharing lessons learned and competition in defining the best solutions.

Once the cooperating partners have decided to develop a mission together, the actual goal and design of the mission can be identified through competition with a call for proposals. The selection of proposals is then, in fact, a competitive process. With competition, the selected mission proposal will ensure the best approach to reach the scientific objectives that have been discussed and agreed upon by the partners.

The next process is to define the division of labor between the cooperating partners. The best way, of course, is to use the best capability that each of the partners can provide. For example, if both partners can provide launchers, the one who can provide the best service takes the upper hand. In general, this is not through competition, but through negotiation. However, the stronger partner with the best services will often have a larger weight in the decision-making.

In the scientific exploitation part, competition exists from the beginning to the end. At the beginning of the cooperation, the scientific group that can propose better ideas for the scientific objectives of the mission, i.e., aiming at the most important frontier that could generate the highest scientific impact, will certainly have a stronger position. In fact, this should not be called competition but rather a part of the cooperation, as both parties will learn each other's strengths. However, when the science data are produced throughout the operational life of the mission, those who make the initial discoveries will certainly demonstrate the team's capabilities, and competition will become evident once again. The publication is generally a joint publication with the first author demonstrating the leadership.

This is not to say that cooperation and competition cannot co-exist. On the contrary, they should stay together all the time. Without cooperation, there will never be a "1+1>2" result, while without competition, the best solution will not emerge during the development of the mission, and there will be no growth for all the science teams involved.

7.6 Responsibility of Science Communities

As explained in the previous chapters, space science missions are mostly government-driven space projects using public funds. In general, scientists are also working in research institutes or universities with public funding. This implies that they should keep pace with government policies when they establish international cooperation. That is to say, if the relations between the governments in question are not ready for the joint development of a space science mission, there will not be fruitful cooperation between the science teams in a joint mission. However, this does not mean that the scientists can do nothing. In the past, we have seen cases demonstrating the ability of scientists to influence governmental policies through their personal efforts.

At the beginning of the space era, how to use outer space was new for all. Scientists helped the governments define the border between air and space, i.e., the 100 km von Kármán line, above the surface of the planet Earth. At the beginning of the 1960s, many scientists called attention to using the outer space peacefully and pushed their governments and the UN to set up the Outer Space Treaty in 1966, which was signed by the major space powers and entered into force in October 1967. Even during the space race, with clear political significance, scientists from both the United States and the Soviet Union worked together to push the two governments to have the famous hand-shake mission in space in 1975, which was a signal of the actual ending of the race.

Right after the US Congress' Wolf Amendment in 2011 which has been blocking any direct contact between NASA and CNSA, scientists from the United States and China gathered together to discuss how to continue relations between the scientific communities on both sides during the 2012 COSPAR assembly. A bilateral forum mechanism for young scientists in space science was set up after that, with the US organizers using private funds.

It is most worth noting that IAA, since its creation in 1960 by von Kármán, never rejected scientists from any country to become a member. Since the beginning, all IAA conferences or symposiums have been open to scientists from all countries regardless of their political backgrounds.

In conclusion, science belongs to all human beings, and due to the global nature of science, scientific communities have the responsibility to think and act with a long-term perspective. Governments have short-term goals and policies, together with financial restrictions. However, scientists can have broader views and stand out for the discovery and exploration of nature for the benefit of humanity as a whole.

References

Bonnet RM, Manno V (1994) International cooperation in space: the example of the European Space Agency. Harvard University Press, Cambridge

Escoubet CP, Fehringer M, Goldstein ML (2001) The cluster mission. Ann Geophys 19(10/12):1197–1200

ESF, NRC (1998) U.S.-European Collaboration in Space Science. National Academy Press, Washington

ESSC-ESF (2000) Future of international collaboration in space science. http://archives.esf.org/fileadmin/Public_documents/Publications/Future_of_International_Collaboration_in_Space_Science.pdf. 20 June 2024

Haerendel G, Torbert RB, Höfner H (1999) The Equator-S mission. Ann Geophys 17(12):1499–1502

Liu ZX, Escoubet P, Cao JB (2005) A Chinese-European multiscale mission: The Double Star Program. Multiscale Coupling Sun-Earth Processes (1):509–514. https://doi.org/10.1016/B978-044451881-1/50037-X

Millan RM, von Steiger R, Ariel M et al (2019) Small satellites for space science: a COSPAR scientific roadmap. Adv Space Res 64(8):1466–1517

Wu J (2022) Introduction to space science. Science Press, Beijing

Wu J, Bonnet R (2017) Maximize the impacts of space science. Nature 551:435–436

Wu J, Giménez A (2020) On the maximization of the science output of space missions. Space Sci Rev 216(1):3

Wu J, Deng L, Praks J et al (2024) CORBES: Radiation belt survey with international small satellite constellation. Adv Space Res. https://doi.org/10.1016/j.asr.2024.04.051. 30 June 2024

© China Science Publishing & Media Ltd. (Science Press) 2025

J. Wu, A. Giménez, *International Cooperation in Space Science*, SpringerBriefs in Applied Sciences and Technology,
https://doi.org/10.1007/978-981-96-8366-6

MIX
Papier aus verantwortungsvollen Quellen
Paper from responsible sources
FSC® C105338

If you have any concerns about our products,
you can contact us on
ProductSafety@springernature.com

In case Publisher is established outside the EU,
the EU authorized representative is:
Springer Nature Customer Service Center GmbH
Europaplatz 3, 69115 Heidelberg, Germany

Printed by Libri Plureos GmbH
in Hamburg, Germany